AF566697

X-Ray Diffraction

A Practical Approach

X-Ray Diffraction

A Practical Approach

C. Suryanarayana
Colorado School of Mines
Golden, Colorado

and

M. Grant Norton
Washington State University
Pullman, Washington

Plenum Press • New York and London

Library of Congress Cataloging in Publication Data

On file

ISBN 0-306-45744-X

© 1998 Plenum Press, New York
A Division of Plenum Publishing Corporation
233 Spring Street, New York, N.Y. 10013

http://www.plenum.com

10 9 8 7 6 5 4 3 2 1

All rights reserved

No part of this book may be reproduced, stored in a retrieval system, or transmitted in any form or by any means, electronic, mechanical, photocopying, microfilming, recording, or otherwise, without written permission from the Publisher

Printed in the United States of America

Preface

X-ray diffraction is an extremely important technique in the field of materials characterization to obtain information on an atomic scale from both crystalline and noncrystalline (amorphous) materials. The discovery of x-ray diffraction by crystals in 1912 (by Max von Laue) and its immediate application to structure determination in 1913 (by W. L. Bragg and his father W. H. Bragg) paved the way for successful utilization of this technique to determine crystal structures of metals and alloys, minerals, inorganic compounds, polymers, and organic materials—in fact, *all* crystalline materials. Subsequently, the technique of x-ray diffraction was also applied to derive information on the fine structure of materials—crystallite size, lattice strain, chemical composition, state of ordering, etc.

Of the numerous available books on x-ray diffraction, most treat the subject on a theoretical basis. Thus, even though you may learn the physics of x-ray diffraction (if you don't get bogged down by the mathematical treatment in some cases), you may have little understanding of how to record an x-ray diffraction pattern and how to derive useful information from it. Thus, the primary aim of this book is to enable students to understand the practical aspects of the technique, analyze x-ray diffraction patterns from a variety of materials under different conditions, and to get the maximum possible information from the diffraction patterns. By doing the experiments using the procedures described herein and following the methods suggested for doing the calculations, you will develop a clear understanding of the subject matter and appreciate how the information obtained can be interpreted.

The book is divided into two parts: *Part I—Basics* and *Part II—Experimental Modules*. Part I covers the fundamental principles necessary to understand the phenomenon of x-ray diffraction. Chapter 1 presents the general background to x-ray diffraction: What are x-rays? How are they produced? How are they diffracted? Chapter 2 reviews the concepts of different types of crystal structures adopted by materials. Additionally, the phenomena of diffraction of x-rays by crystalline materials, concepts of structure factor, and selection rules for the observance (or absence) of reflections are explained. Chapter 3 presents an overview of the experimental considerations involved in obtaining useful x-ray diffraction patterns and a brief introduction to the interpretation and significance of x-ray diffraction patterns. Even though the theoretical aspects are discussed in Part I, we have adopted an approach quite different from that of other textbooks in that we lay more emphasis on the physical significance of the phenomenon and concepts rather than burden you with heavy mathematics. We have used boxed text to further explain some particular, or possibly confusing, aspects.

Part II contains eight experimental modules. Each module covers one topic. For example, the first module explains how an x-ray diffraction pattern obtained from a cubic material can be indexed. First we go through the necessary theory, using the minimum amount of mathematics. Then we do a worked example based on actual experimental data we have obtained; this is followed by an experiment for you to do. Finally we have included a few exercises based on the content of the module. These give you a chance to apply further some of the knowledge you have acquired. Each experimental module follows a similar format. We have also made each module self-contained; so you can work through them in any order, however, we suggest you do Experimental Module 1 first since this provides a lot of important background information which you may find useful when you work through some of the later modules. By working through the modules, or at least a selection of them, you will discover what information can be obtained by x-ray diffraction and, more importantly, how to interpret that information. Work tables have been provided so that you can tabulate your data and results. Further, we have taken examples from *all* categories of materials—metals, ceramics, semiconductors, and polymers—to emphasize that x-ray diffraction can be effectively and elegantly used to characterize any type of material. This is an important feature of our approach.

Another important feature of the book is that it provides x-ray diffraction patterns for all the experiments and lists the values of the Bragg angles (diffraction angles, θ). Therefore, even if you have no access to an x-ray diffractometer, or if the unit is down, you can use these 2θ values and perform the calculations. Alternatively, if you are able to record the x-ray diffraction patterns, the patterns provided in the book can be used as a reference; you can compare the pattern you recorded against what is given in the book.

This book is primarily intended for use by undergraduate junior or senior-level students majoring in materials science or metallurgy. However, the book can also be used very effectively by undergraduate students of geology, physics, chemistry, or any other physical science likely to use the technique of x-ray diffraction for materials characterization. Preliminary knowledge of freshman physics and simple ideas of crystallography will be useful but not essential because these have been explained in easy-to-understand terms in Part I.

The eight modules in Part II can be easily completed in a one-semester course on x-ray diffraction. If x-ray diffraction forms only a part of a broader course on materials characterization, then not all the modules need to be completed.

We realize that we have not included *all* possible applications of x-ray diffraction to materials. The book deals only with polycrystalline materials (mostly powders). We are aware that there are other important applications of x-ray diffraction to polycrystalline materials. Since this book is intended for an undergraduate course, and some special and advanced topics are not covered in most undergraduate programs, we have not discussed topics such as stress measurement and texture analysis in polycrystalline materials. X-ray diffraction can also be used to obtain structural information about single crystals and their orientation and the structure of noncrystalline (amorphous) materials. But this requires use of a slightly different experimental setup or sophisticated software which is not available in most undergraduate laboratories. For this reason we have not covered these topics.

Pullman, WA

C. Suryanarayana
M. Grant Norton

Acknowledgments

In writing any book it is unlikely that the authors have worked entirely in isolation without assistance from colleagues and friends. We are certainly not exceptions and it is with great pleasure that we acknowledge those people that have contributed, in various ways, to this project.

We are indebted to Mr. Enhong Zhou and Mr. Charles Knowles of the University of Idaho for helping to record all the x-ray diffraction patterns in this book. Their attention to detail and their flexibility in accommodating our schedule are gratefully appreciated. The entire manuscript was read by Professor John Hirth, Professor Kelly Miller, and Mr. Sreekantham Sreevatsa, and we thank them for their time and effort and their helpful suggestions and comments. Dr. Frank McClune of the International Centre for Diffraction Data provided us with the latest information from the Powder Diffraction File. Dr. Vinod Sikka of Oak Ridge National Laboratory generously provided us with an ingot of Cu_3Au. Simon Bates of Philips Analytical X-Ray, Rick Smith of Osmic, Inc., and David Aloisi of X-Ray Optical Systems, Inc., contributed helpful discussions and information on recent developments in x-ray instrumentation. This book was written while one of the authors (CS) was a Visiting Professor at Washington State University in Pullman. We are both obliged to Professor Stephen Antolovich, Director of the School of Mechanical and Materials Engineering at Washington State University, for facilitating our collaboration and for providing an environment wherein we could complete this book.

And last, but by no means least, we would like to thank our wives Meena and Christine. Their presence provides us with an invisible staff that makes the journey easier. It is to them that we dedicate this book.

Contents

Part I. Basics

Part I

Basics

1

X-Rays and Diffraction

1.1. X-RAYS

X-rays are high-energy electromagnetic radiation. They have energies ranging from about 200 eV to 1 MeV, which puts them between γ-rays and ultraviolet (UV) radiation in the electromagnetic spectrum. It is important to realize that there are no sharp boundaries between different regions of the electromagnetic spectrum and that the assigned boundaries between regions are arbitrary. Gamma rays and x-rays are essentially

The Electron Volt

Materials scientists and physicists often use the electron volt (eV) as the unit of energy. An electron volt is the amount of energy an electron picks up when it moves between a potential (voltage) difference of 1 volt. Thus,

$$1 \text{ eV} = 1.602 \times 10^{-19} \text{ C (the charge on an electron)} \times 1 \text{ V} = 1.602 \times 10^{-19} \text{ J}$$

Although the eV has been superseded by the joule (J)—the SI unit of energy—the eV is a very convenient unit when atomic-level processes are being represented. For example, the ground-state energy of an electron in a hydrogen atom is –13.6 eV; to form a vacancy in an aluminum crystal requires 0.76 eV. The eV is used almost exclusively to represent electron energies in electron microscopy. The conversion factor between eV and J is 1 eV = 1.602×10^{-19} J. Most texts on materials characterization techniques use the electron volt, so you should familiarize yourself with this unit.

The Ångstrom

The ångstrom (Å) is a unit of length equal to 10^{-10} m. The ångstrom was widely used as a unit of wavelength for electromagnetic radiation covering the visible part of the electromagnetic spectrum and x-rays. This unit is also used for interatomic spacings, since these distances then have single-digit values. Although the ångstrom has been superseded in SI units by the nanometer (1 nm = 10^{-9} m = 10 Å), many crystallographers and microscopists still prefer the older unit. Once again, it is necessary for you to become familiar with both units. Throughout this text (except in Experimental Module 8) we use the nanometer.

identical, γ-rays being somewhat more energetic and shorter in wavelength than x-rays. Gamma-rays and x-rays differ mainly in how they are produced in the atom. As we shall see presently, x-rays are produced by interactions between an external beam of electrons and the electrons in

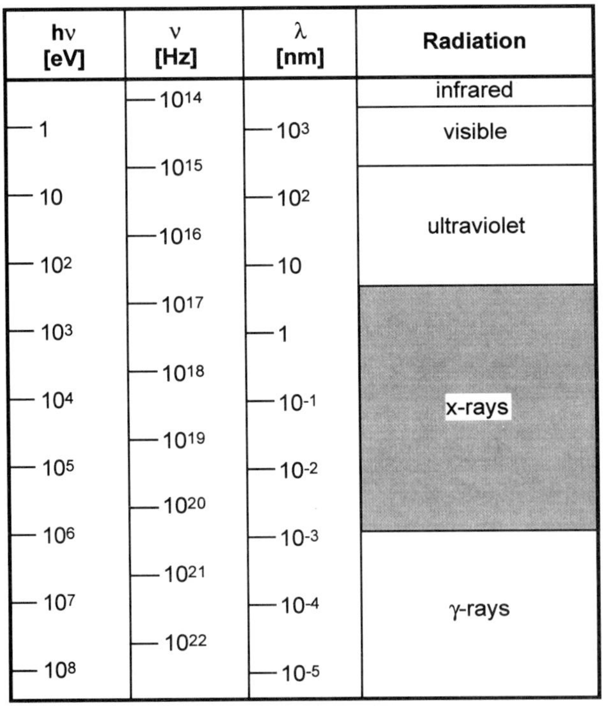

hν [eV]	ν [Hz]	λ [nm]	Radiation
	10^{14}		infrared
1		10^{3}	visible
	10^{15}		
10		10^{2}	
	10^{16}		ultraviolet
10^{2}		10	
	10^{17}		
10^{3}		1	
	10^{18}		
10^{4}		10^{-1}	x-rays
	10^{19}		
10^{5}		10^{-2}	
	10^{20}		
10^{6}		10^{-3}	
	10^{21}		
10^{7}		10^{-4}	γ-rays
	10^{22}		
10^{8}		10^{-5}	

FIG. 1. Part of the electromagnetic spectrum. Note that the boundaries between regions are arbitrary. The usable range of x-ray wavelengths for x-ray diffraction studies is between 0.05 and 0.25 nm (only a small part of the total range of x-ray wavelengths).

the shells of an atom. On the other hand, γ-rays are produced by changes within the nucleus of the atom. A part of the electromagnetic spectrum is shown in Fig. 1.

Each quantum of electromagnetic radiation, or *photon*, has an energy, E, which is proportional to its frequency, ν:

$$E = h\nu \tag{1}$$

The constant of proportionality is Planck's constant h, which has a value of 4.136×10^{-15} eV·s (or 6.626×10^{-34} J·s). Since the frequency is related to the wavelength, λ, through the speed of light, c, the wavelength of the x-rays can be written as

$$\lambda = \frac{hc}{E} \tag{2}$$

where c is 2.998×10^{8} m/s. So, using the energies given at the beginning of this section, we can see that x-ray wavelengths vary from about 10 nm to 1 pm. Notice that the wavelength is shorter for higher energies. The useful range of wavelengths for x-ray diffraction studies is between 0.05 and 0.25 nm. You may recall that interatomic spacings in crystals are typically about 0.2 nm (2 Å).

1.2. THE PRODUCTION OF X-RAYS

X-rays are produced in an x-ray tube consisting of two metal electrodes enclosed in a vacuum chamber, as shown in cross section in Fig. 2. Electrons are produced by heating a tungsten filament cathode. The cathode is at a high negative potential, and the electrons are accelerated toward the anode, which is normally at ground potential. The electrons, which have a very high velocity, collide with the water-cooled anode. The loss of energy of the electrons due to the impact with the metal anode is manifested as x-rays. Actually only a small percentage (less than 1%) of the electron beam is converted to x-rays; the majority is dissipated as heat in the water-cooled metal anode.

A typical x-ray spectrum, in this case for molybdenum, is shown in Fig. 3. As you can see, the spectrum consists of a range of wavelengths. For each accelerating potential a continuous x-ray spectrum (also known as the white spectrum), made up of many different wavelengths, is obtained. The continuous spectrum is due to electrons losing their energy in a series of collisions with the atoms that make up the target, as shown in Fig. 4. Because each electron loses its energy in a different way, a continuous spectrum of energies and, hence, x-ray wavelengths is pro-

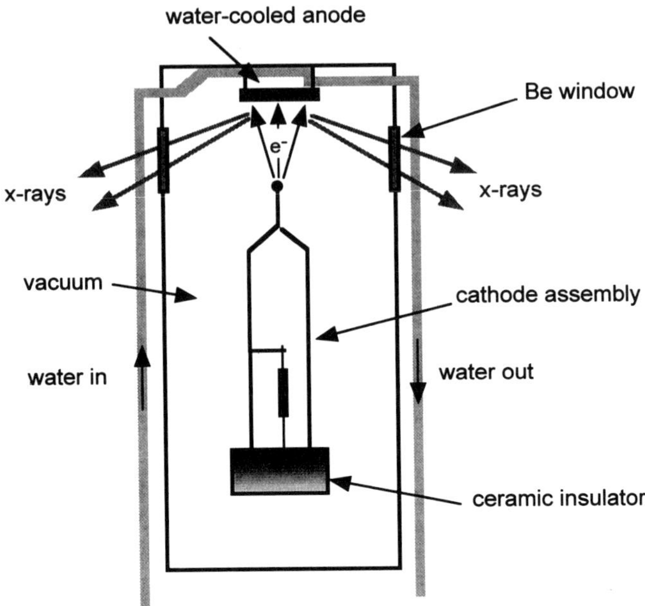

FIG. 2. Schematic showing the essential components of a modern x-ray tube. Beryllium is used for the window because it is highly transparent to x-rays.

duced. We don't normally use the continuous part of the x-ray spectrum unless we require a number of different wavelengths in an experiment, for example in the Laue method (which we will not describe).

If an electron loses all its energy in a single collision with a target atom, an x-ray photon with the maximum energy or the shortest wavelength is produced. This wavelength is known as the short-wavelength limit (λ_{SWL}) and is indicated in Fig. 3 for a molybdenum target irradiated with 25-keV electrons. [Note: When referring to electron energies, we use either eV or keV, but when referring to the accelerating potential applied to the electron, we use V or kV.]

If the incident electron has sufficient energy to eject an inner-shell electron, the atom will be left in an excited state with a hole in the electron shell. This process is illustrated in Fig. 5. When this hole is filled by an electron from an outer shell, an x-ray photon with an energy equal to the difference in the electron energy levels is produced. The energy of the x-ray photon is characteristic of the target metal. The sharp peaks, called *characteristic lines,* are superimposed on the continuous spectrum, as shown in Fig. 3. It is these characteristic lines that are most useful in x-ray diffraction work, and we deal with these later in the book.

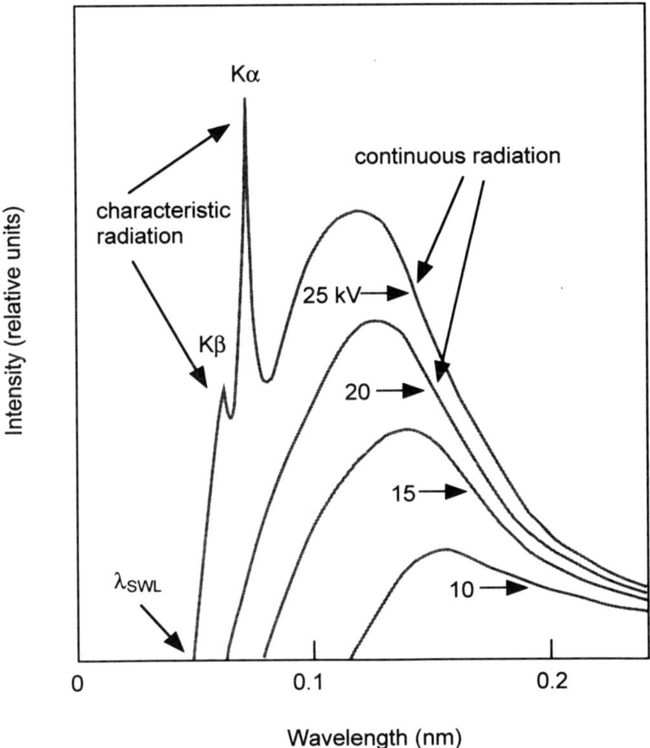

FIG. 3. X-ray spectrum of molybdenum at different potentials. The potentials refer to those applied between the anode and cathode. (The linewidths of the characteristic radiation are not to scale.)

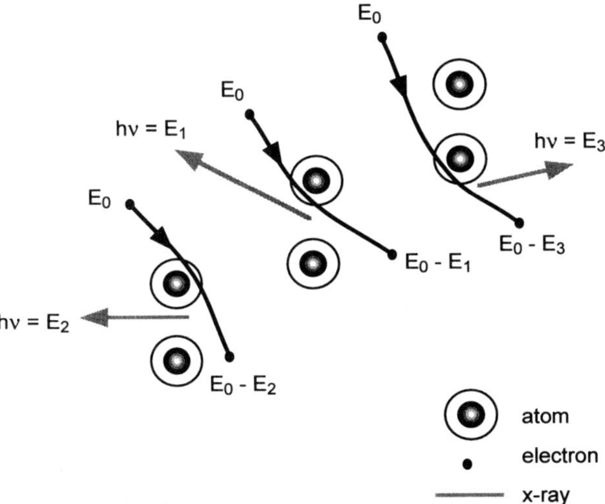

FIG. 4. Illustration of the origin of continuous radiation in the x-ray spectrum. Each electron, with initial energy E_0, loses some, or all, of its energy through collisions with atoms in the target. The energy of the emitted photon is equal to the energy lost in the collision.

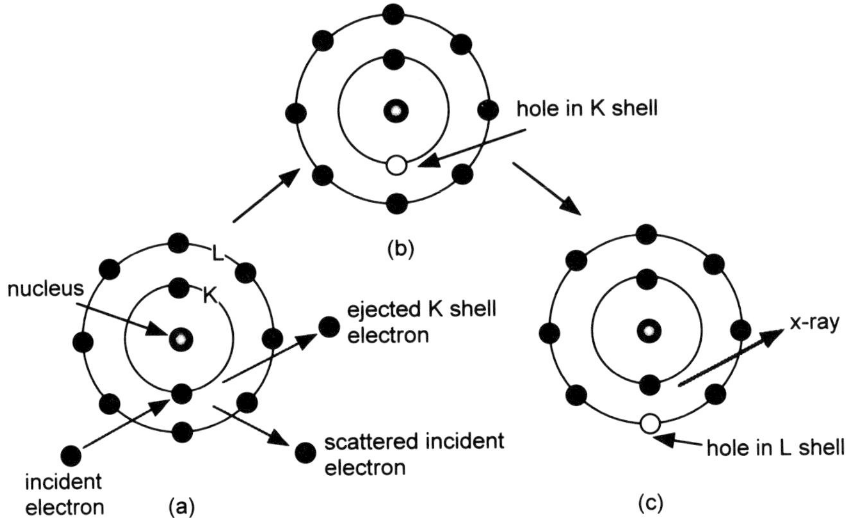

FIG. 5. Illustration of the process of inner-shell ionization and the subsequent emission of a characteristic x-ray: (a) an incident electron ejects a K shell electron from an atom, (b) leaving a hole in the K shell; (c) electron rearrangement occurs, resulting in the emission of an x-ray photon.

If the entire electron energy is converted to that of the x-ray photon, the energy of the x-ray photon is related to the excitation potential V experienced by the electron:

$$E = \frac{hc}{\lambda} = eV \tag{3}$$

where e is the electron charge (1.602×10^{-19} C). The x-ray wavelength is thus

$$\lambda = \frac{hc}{eV} \tag{4}$$

Inserting the values of the constants h, c, and e, we have

$$\lambda\ [\text{nm}] = \frac{1.243}{V} \tag{5}$$

when the potential V is expressed in kV. This wavelength corresponds to λ_{SWL}; the characteristic lines will have wavelengths longer than λ_{SWL}. The accelerating potentials necessary to produce x-rays having wavelengths comparable to interatomic spacings are therefore about 10 kV. Higher accelerating potentials are normally used to produce a higher-intensity line spectrum characteristic of the target metal. The use of higher accelerating potentials changes the value of λ_{SWL} but not the characteristic wavelengths. The intensity of a characteristic line depends on both the applied potential and the tube current i (the number of electrons per second striking the target). For an applied potential V, the intensity of the K lines shown in Fig. 3 is approximately

$$I = Bi(V - V_K)^n \tag{6}$$

where B is a proportionality constant, V_K is the potential required to eject an electron from the K shell, and n is a constant, for a particular value of V, which has a value between 1 and 2.

As you can see in Fig. 3, there is more than one characteristic line. The different characteristic lines correspond to electron transitions between different energy levels. The characteristic lines are classified as K, L, M, etc. This terminology is related to the Bohr model of the atom in which the electrons are pictured as orbiting the nucleus in specific shells. For historical reasons, the innermost shell of electrons is called the K shell, the next innermost one the L shell, the next one the M shell, and so on.

If we fill a hole in the K shell with an electron from the L shell, we get a Kα x-ray, but if we fill the hole with an electron from the M shell, we get a Kβ x-ray. If the hole is in the L shell and we fill it with an electron from the M shell we get an Lα x-ray. Figure 6 shows schematically the origin of these three different characteristic lines.

The situation is complicated by the presence of subshells. For example, we differentiate the Kα x-rays as $K\alpha_1$ and $K\alpha_2$. The reason for this differentiation is that the L shell consists of three subshells, L_I, L_{II}, and L_{III}; a transition from L_{III} to K results in emission of the $K\alpha_1$ x-ray and a

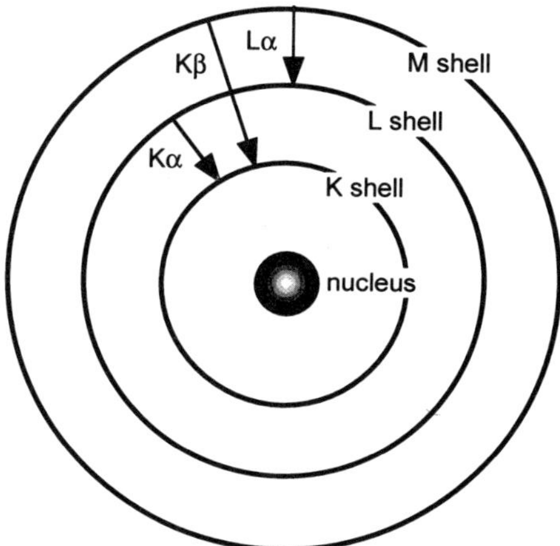

FIG. 6. Electron transitions in an atom, which produce the Kα, Kβ, and Lα characteristic x-rays.

Quantum Numbers

You are probably familiar with assigning quantum numbers to the electrons in an atom and writing down the electron configuration of an atom based on these quantum numbers. For example, the electron configuration of silicon (Si), atomic number 14, is $1s^2 2s^2 2p^6 3s^2 3p^2$. The first number is the value of the principal quantum number n. For the K shell, $n = 1$, for the L shell $n = 2$, for the M shell $n = 3$, and so on. The letter (s, p, etc.) represents the value of the orbital-shape quantum number, l. For the K shell there are no subshells because there is only one value of l; $l = 0$. For the L shell there are subshells because there are two values of l; $l = 0$ and $l = 1$. These values of l correspond to the $2s$ and the $2p$ levels, respectively.

transition from L_{II} to K results in emission of the $K\alpha_2$ x-ray. All the shells except the K shell have subshells.

Let's do an example to illustrate these different transitions for molybdenum. The energies of the K, L_{II}, and L_{III} levels are given in Table 1. The wavelength of the emitted x-rays is related to the energy difference between any two levels by Eq. (2). The energy difference between the L_{III} and K levels is 17.48 keV. Using this energy in Eq. (2) and substituting in

Designation of Subshells and Angular Momentum

We now introduce a new quantum number, j, which represents the total angular momentum of an electron:

$$j = l + m_s$$

where m_s is the spin quantum number, which, you may remember, can have values of ±1/2. The values of j can only be positive numbers, so for the L shell we obtain

Subshell notation	n	l	m_s	j
L_I	2	0	$+\frac{1}{2}$	$\frac{1}{2}$
L_{II}	2	1	$-\frac{1}{2}$	$\frac{1}{2}$
L_{III}	2	1	$+\frac{1}{2}$	$\frac{3}{2}$

It is the presence of these subshells that gives rise to splitting of the characteristic lines in the x-ray spectrum.

TABLE 1. Energies of the K, L_{II}, and L_{III} Levels of Molybdenum

Level	Energy (keV)
K	–20.00
L_{II}	–2.63
L_{III}	–2.52

the constants, we obtain a wavelength of λ = 0.0709 nm. This is the wavelength of the Kα_1 x-rays of Mo. The energy difference between the L_{II} and K levels is 17.37 keV. Using Eq. (2) again, we obtain the wavelength λ = 0.0714 nm. This is the wavelength of the Kα_2 x-rays of Mo.

Figure 7 shows the x-ray spectrum for Mo at 35 kV. The right-hand-side figure shows the well-resolved Kα doublet on an expanded energy (wavelength) scale. However, it is not always possible to resolve (separate) the Kα_1 and Kα_2 lines in the x-ray spectrum because their wavelengths are so close. If the Kα_1 and Kα_2 lines cannot be resolved, the characteristic line is simply called the Kα line and the wavelength is given by the weighted average of the Kα_1 and Kα_2 lines.

Figure 8 shows the complete range of allowed electron transitions in a molybdenum atom. Not all the electron transitions are equally probable. For example, the Kα transition (i.e., an electron from the L shell filling a hole in the K shell) is 10 times more likely than the Kβ transition (i.e., an electron from the M shell filling a hole in the K shell).

Weighted Average

Sometimes it is not possible to resolve the Kα_1 and Kα_2 lines in the x-ray spectrum. In these cases we take the wavelength of the unresolved Kα line as the weighted average of the wavelengths of its components. To determine the weighted average, we need to know not only the wavelengths of the resolved lines but also their relative intensities. The Kα_1 line is twice as strong (intense) as the Kα_2 line, so it is given twice the weight. The wavelength of the unresolved Mo Kα line is thus

$$\frac{1}{3}(2 \times 0.0709 + 0.0714) = 0.0711 \text{ nm}$$

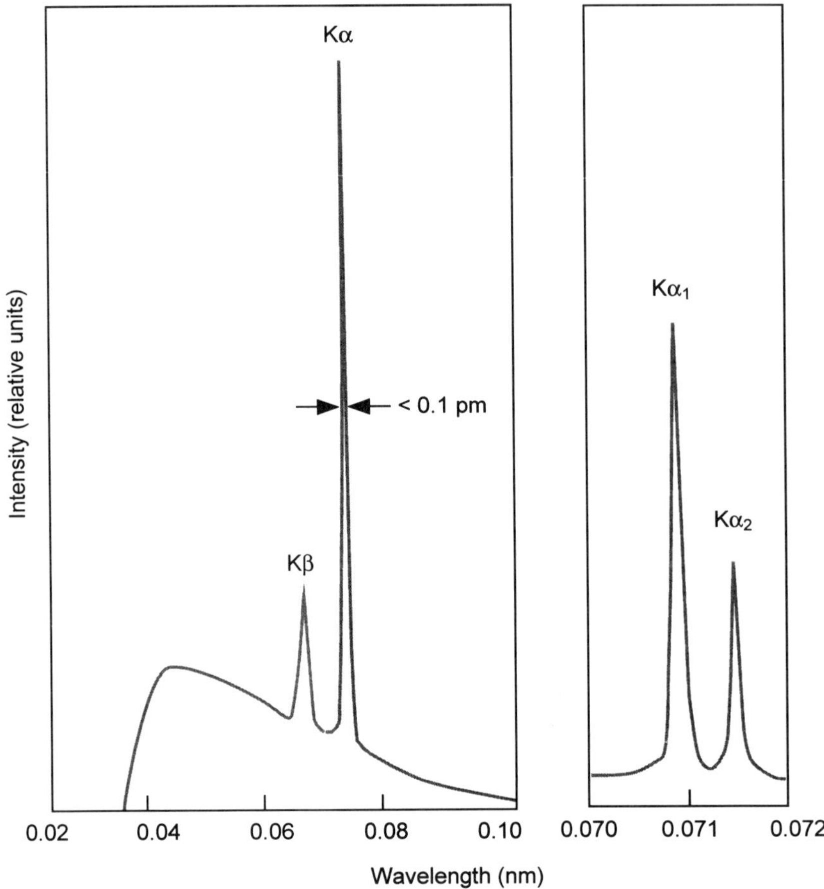

FIG. 7. X-ray spectrum of molybdenum at 35 kV. The expanded scale on the right shows the resolved Kα_1 and Kα_2 lines.

The important radiations in diffraction work are those corresponding to the filling of the innermost K shell from adjacent shells giving the so-called Kα_1, Kα_2, and Kβ lines. For copper, molybdenum, and some other commonly used x-ray sources, the characteristic wavelengths to six decimal places are given in Table 2.

For most x-ray diffraction studies we want to use a *monochromatic* beam (x-rays of a single wavelength). The simplest way to obtain this is to filter out the unwanted x-ray lines by using a foil of a suitable metal whose absorption edge for x-rays lies between the Kα and Kβ components of the spectrum. The absorption edge, or, as it is also known, critical absorption wavelength represents an abrupt change in the absorption characteristics of x-rays of a particular wavelength by a material. For example, a nickel

Selection Rules Governing Electron Transitions

In Fig. 8 and in the preceding discussion you may have noticed that there is no electron transition between the L_I subshell and the K shell. The reason for this, and the absence of other transitions, is based on a series of selection rules governing electron transitions. A detailed description of why these transitions are absent would require us to discuss the Schrödinger wave equation (the famous equation that relates the wavelike properties of an electron to its energy), which is beyond the scope of this book. But we can use the results that come from the Schrödinger equation, which show that the selection rules for electron transitions are

$$\Delta n = \text{anything}$$

$$\Delta l = \pm 1$$

$$\Delta j = 0 \text{ or } \pm 1$$

where Δn is the change in the principal quantum number, Δl is the change in the orbital-shape quantum number, and Δj is the change in the angular-momentum quantum number. Transitions between any shell (principal quantum number) are allowed (e.g., $2p \rightarrow 1s$), but transitions where the change in l is zero are not allowed (e.g., $2s \rightarrow 1s$). Therefore the L_I to K transition is not allowed.

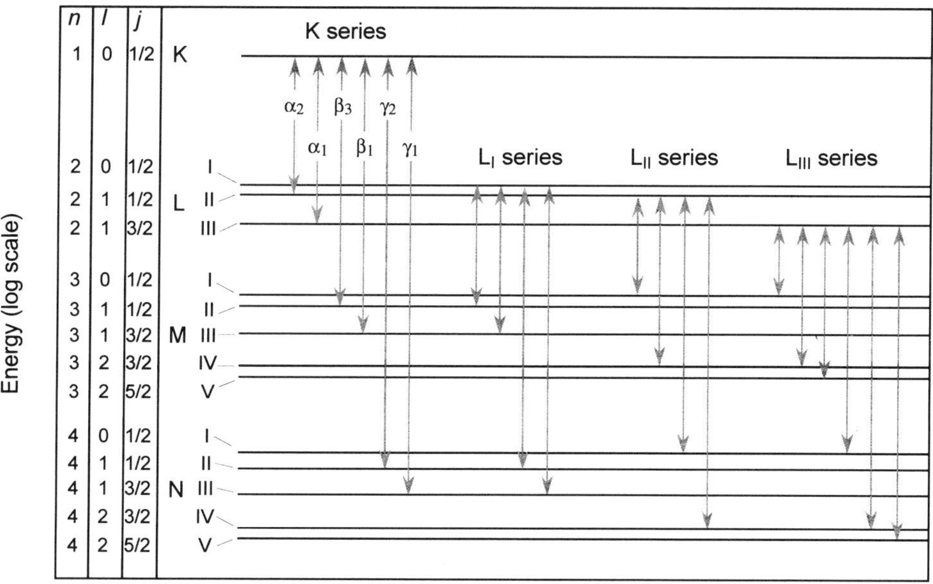

FIG. 8. Energy-level diagram showing all the allowed electron transitions in a molybdenum atom.

TABLE 2. Some Commonly Used X-Ray K Wavelengths (in nm)

Element	Kα (weighted average)	$K\alpha_2$ (strong)	$K\alpha_1$ (very strong)	Kβ (weak)
Cr	0.229100	0.229361	0.228970	0.208487
Fe	0.193736	0.193998	0.193604	0.175661
Co	0.179026	0.179285	0.178897	0.162079
Cu	0.154184	0.154439	0.154056	0.139222
Mo	0.071073	0.071359	0.070930	0.063229

foil will remove Cu Kβ radiation, and zirconium will remove Mo Kβ radiation. However, in most modern x-ray diffractometers a monochromatic beam is obtained by using a crystal monochromator. A crystal monochromator consists of a crystal, graphite is often used, with a known lattice spacing oriented in such a way that it only diffracts the Kα radiation, and not the Kβ radiation. The beam is still made up of the $K\alpha_1$ and $K\alpha_2$ wavelengths.

For x-ray diffraction studies there is a wide choice of characteristic Kα lines obtained by using different target metals, as shown in Table 2, but, Cu Kα is the most common radiation used. The Kα lines are used because they are more energetic than Lα and therefore less strongly absorbed by the material we want to examine. The wavelength spread of each line is extremely narrow, and each wavelength is known with very high precision.

1.3. DIFFRACTION

Diffraction is a general characteristic of all waves and can be defined as the modification of the behavior of light or other waves by its interaction with an object. You should already be familiar with the term "diffraction" from introductory physics classes. In this section we review some fundamental features of diffraction, particularly as they apply to the use of x-rays for determining crystal structures.

First let's consider an individual isolated atom. If a beam of x-rays is incident on the atom, the electrons in the atom then oscillate about their mean positions. Recall from Section 1.2 that when an electron decelerates (loses energy) it emits x-rays. This process of absorption and reemission of electromagnetic radiation is known as *scattering*. Using the concept of a photon, we can say that an x-ray photon is absorbed by the atom and another photon of the same energy is emitted. When there is no change in energy between the incident photon and the emitted photon, we say that the radiation has been *elastically* scattered. On the other hand, inelastic scattering involves photon energy loss.

If the atom we choose to consider is anything other than hydrogen, we would have to consider scattering from more than one electron. Figure 9 shows an atom containing several electrons arranged as points around the nucleus. Although you know from quantum mechanics that this is not a correct representation of atomic structure, it helps our explanation. We are concerned with what happens to two waves that are incident on the atom. The upper wave is scattered by electron A in the forward direction. The lower wave is scattered in the forward direction by electron B. The two waves scattered in the forward direction are said to be in phase (in step or coherent are other terms we use) across wavefront XX′ since these waves have traveled the same total distance before and after scattering; in other words, there is no path (or phase) difference. (A wavefront is simply a surface perpendicular to the direction of propagation of the wave.) If the two waves are in phase, then the maximum in one wave is aligned with the maximum in the other wave. If we add these two waves across wavefront XX′ (i.e., we sum their amplitudes), we obtain a wave with the same wavelength but twice the amplitude.

The other scattered waves in Fig. 9 will not be in phase across wavefront YY′ when the path difference (CB – AD) is not an integral number of wavelengths. If we add these two waves across wavefront YY′, we find that the amplitude of the scattered wave is less than the amplitude of the wave scattered by the same electrons in the forward direction.

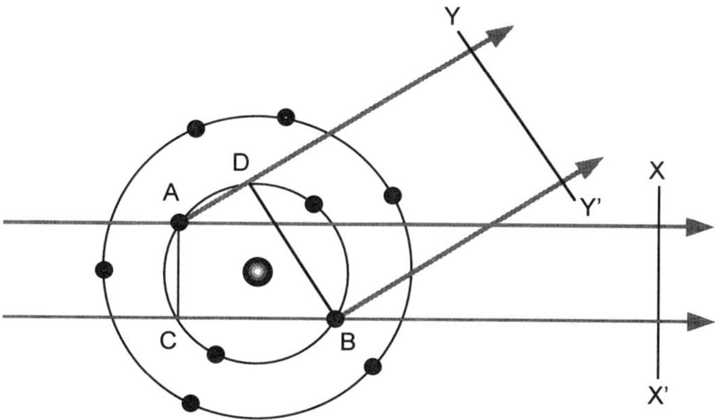

FIG. 9. Scattering of x-rays by an atom.

The Superposition of Waves

When two waves are moving through the same region of space they will superimpose (overlap). The resultant wave is the algebraic sum of the various amplitudes at each point. This is known as the superposition principle. Figure 10 shows three examples of the superposition of two waves: (a) when the component waves are in phase, we have constructive interference and the resultant wave amplitude is large; (b) when the phase difference increases, the amplitude of the resultant wave decreases; and (c) when the component waves are 180° out of phase, the resultant wave has its smallest amplitude and we have destructive interference. Since the amplitudes of wave 1 and wave 2 are different, there is some resultant amplitude. If the amplitudes of waves 1 and 2 are equal, then the resultant amplitude is zero and there is no intensity.

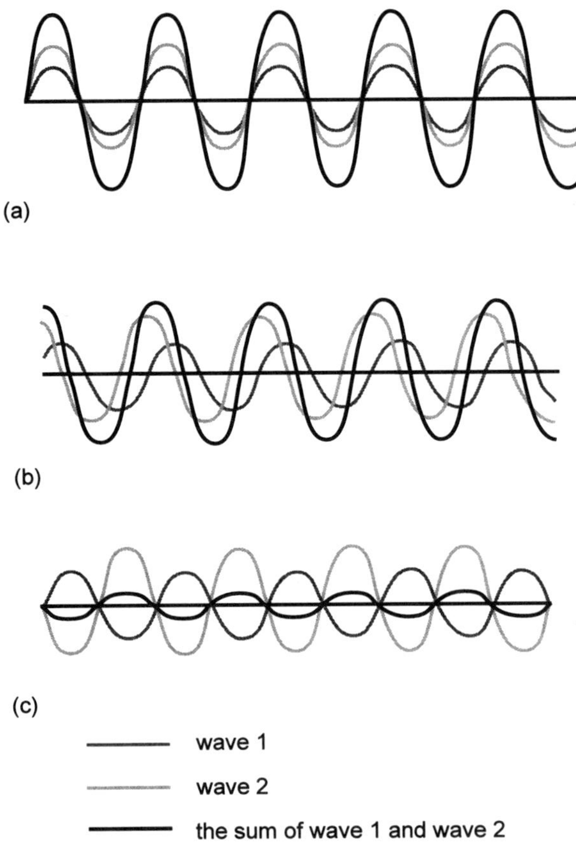

FIG. 10. Illustration of the superposition of waves.

We define a quantity called the *atomic scattering factor*, f, to explain how efficient an atom is scattering in a given direction:

$$f = \frac{\text{Amplitude of wave scattered by an atom}}{\text{Amplitude of wave scattered by one electron}} \tag{7}$$

When scattering is in the forward direction (i.e., the scattering angle, $\theta = 0°$) $f = Z$ (the atomic number—i.e., the total number of electrons) since the waves scattered by all the electrons in the atom are in phase and the amplitudes sum up. But as θ increases, the waves become more and more out of phase because they travel different path lengths and, therefore, the amplitude, or f, decreases. The atomic scattering factor also depends on the wavelength of the incident x-rays. For a fixed value of θ, f is smaller for shorter-wavelength radiation. The variation of atomic scattering factor with scattering angle for copper, aluminum, and oxygen, is shown in Fig. 11. The curves begin at the atomic number (Z), which for copper is 29, and decrease with increasing values of θ or decreasing values of λ. In fact, f is generally plotted against $(\sin\theta)/\lambda$ to take into account the variation of f with both θ and λ. The rate of decrease of f with

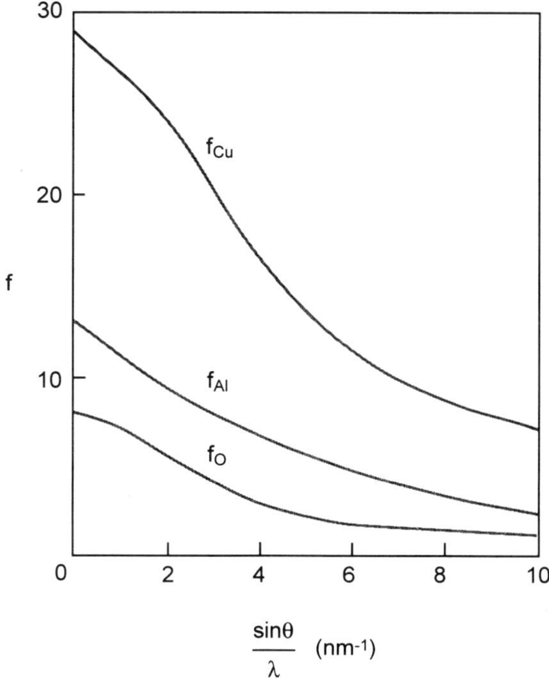

FIG. 11. Variation of the atomic scattering factor of copper, aluminum, and oxygen with $(\sin\theta)/\lambda$.

increasing values of (sin θ)/λ is different for different elements, as you can see in Fig. 11. Note that most of the scattering occurs in the forward direction, when $\theta \approx 0°$.

Let's now consider some closely spaced atoms each of which contributes many scattered x-rays. The scattered waves from each atom interfere. If the waves are in phase, then *constructive interference* occurs. If the waves are 180° out of phase, then *destructive interference* occurs. A diffracted beam may be defined as a beam composed of a large number of superimposed scattered waves. For a measurable diffracted beam complete destructive interference does not occur.

To describe diffraction we have introduced three terms:

- Scattering
- Interference
- Diffraction

What is the difference among these terms? Scattering is the process whereby the incident radiation is absorbed and then reemitted in different directions. Interference is the superposition of two or more of these scattered waves, producing a resultant wave that is the sum of the overlapping wave contributions. Diffraction is constructive interference of more than one scattered wave. There is no real physical difference between constructive interference and diffraction.

1.4. A VERY BRIEF HISTORICAL PERSPECTIVE

If we look back into history (hindsight is a great thing!), the first inkling that diffraction may be useful for studying crystal structure came from the classic double-slit experiment performed by Thomas Young over 200 years ago. At the time Young may well not have realized that the phenomenon he observed would have application to other forms of electromagnetic radiation, and certainly he was not aware of x-rays. Young died in 1829, sixty-six years before the discovery of x-rays by Wilhelm Röntgen in 1895.

In Young's double-slit experiment two coherent (i.e., in phase) beams of light obtained by passing light through two parallel slits were allowed to interfere. The pattern produced on a screen placed beyond the slits consisted of a series of bright and dark lines, as shown schematically in Fig. 12. If we replace the double slits with a grid consisting of many parallel slits, called a diffraction grating, and shine a line source of electromagnetic radiation on the grid, we also observe a pattern consisting of a series of bright and dark lines. The separation of the lines depends on the wavelength (λ) of the radiation and the spacing (d) between the slits in the

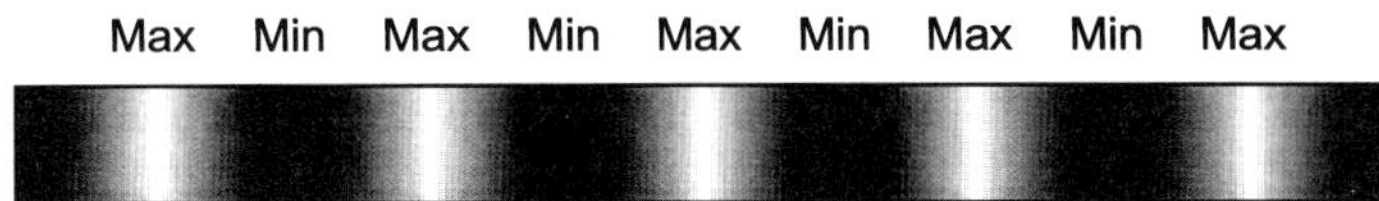

FIG. 12. The fringe pattern produced on a screen in Young's experiment. Waves passing through two slits interfere, and the pattern observed on the screen consists of a series of white (max) and dark (min) lines (not drawn to scale.)

grating. If two diffraction gratings are now superimposed with their lines intersecting at right angles (like a possible arrangement of the lattice planes in a crystal), a spot pattern is produced in which the distance between the spots is a function of the spacing in the gratings and the wavelength of the radiation for a given pair of diffraction gratings. For the experiment to work, the dimensions of the slits in the grating must be comparable to the wavelength of the radiation used.

Max von Laue, in 1912, realized that if x-rays had a wavelength similar to the spacing of atomic planes in a crystal, then it should be possible to diffract x-rays by a crystal and, hence, to obtain information about the arrangement of atoms in crystals.

2

Lattices and Crystal Structures

2.1. TYPES OF SOLID AND ORDER

We can classify solids into three general categories:

1. Single crystal
2. Polycrystalline
3. Amorphous

Simple schematics representing these three categories are shown in Fig. 13.

A crystal is said to possess long-range order because it is composed of atoms arranged in a regular ordered pattern in three dimensions. This periodic arrangement, known as the *crystal structure,* extends over dis-

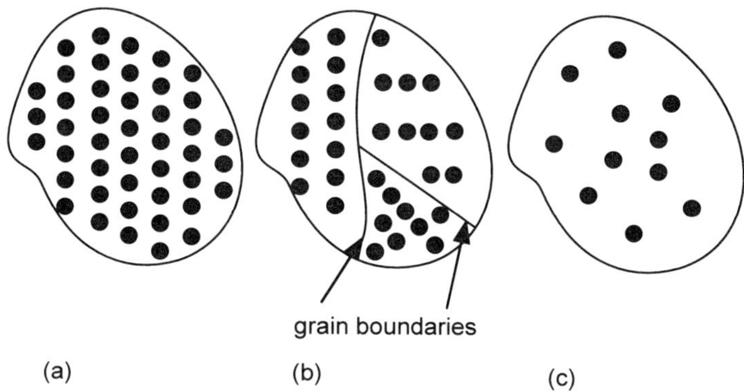

FIG. 13. Illustration of the difference between (a) single crystal, (b) polycrystalline, and (c) amorphous materials.

Grain Boundaries

Grains in a polycrystalline material are generally in many different orientations. The boundary between the grains—the grain boundary—depends on the misorientation of the two grains and the rotation axis about which the misorientation has occurred. There are two special types of grain boundary, illustrated in Fig. 14, which are relatively simple to visualize: the pure tilt boundary and the pure twist boundary. In a tilt boundary the axis of rotation is parallel to the grain-boundary plane. In a twist boundary, the rotation axis is perpendicular to the grain-boundary plane. In general, the axis of rotation will not be simply oriented with respect to either the grain or the grain-boundary plane.

tances much larger than the interatomic separation. (Remember, interatomic separations are about 0.2 nm.) In a single crystal this order extends throughout the entire volume of the material.

A polycrystalline material consists of many small single-crystal regions (called grains) separated by grain boundaries. The grains on either side of the grain boundary are misoriented with respect to each other. The grains in a polycrystalline material can have different shapes and sizes.

In amorphous materials, such as glasses and many polymers, the atoms are not arranged in a regular periodic manner. *Amorphous* is a Greek word meaning "without definite form." Amorphous materials possess only short-range order. The order only extends to a few of the nearest neighbors—distances of less than a nanometer. (A glass is not really a solid; it is actually a supercooled liquid with a very high viscosity—about 15 orders

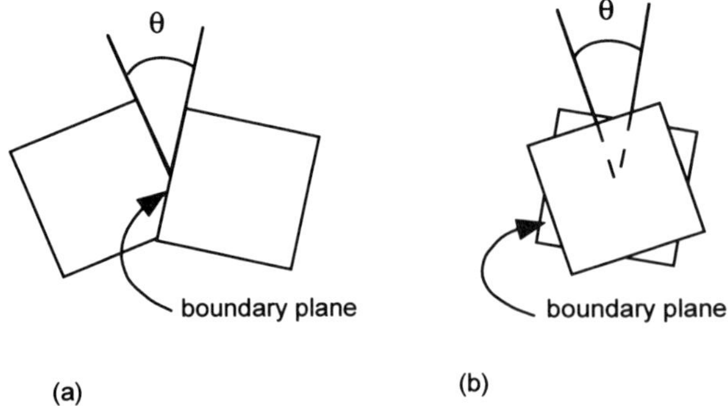

FIG. 14. Illustration of (a) a tilt grain boundary and (b) a twist grain boundary.

of magnitude greater than water at room temperature—so that in many respects it behaves like a solid.)

2.2. POINT LATTICES AND THE UNIT CELL

Let's consider the three-dimensional arrangement of points in Fig. 15. This arrangement is called a *point lattice*. If we take any point in the point lattice it has exactly the same number and arrangement of neighbors (i.e., identical surroundings) as any other point in the lattice. This condition should be fairly obvious considering our description of long-range order in Sec. 2.1. We can also see from Fig. 15 that it is possible to divide the point lattice into much smaller units such that when these units are stacked in three dimensions they reproduce the point lattice. This small repeating unit is known as the *unit cell* of the lattice and is shown in Fig. 16.

A unit cell may be described by the interrelationship between the lengths (a, b, c) of its sides and the interaxial angles (α, β, γ) between them. (α is the angle between the b and c axes, β is the angle between the a and c axes, and γ is the angle between the a and b axes.) The actual values of a, b, and c, and α, β, and γ are not important, but their interrelation is. The lengths are measured from one corner of the cell, which is taken as

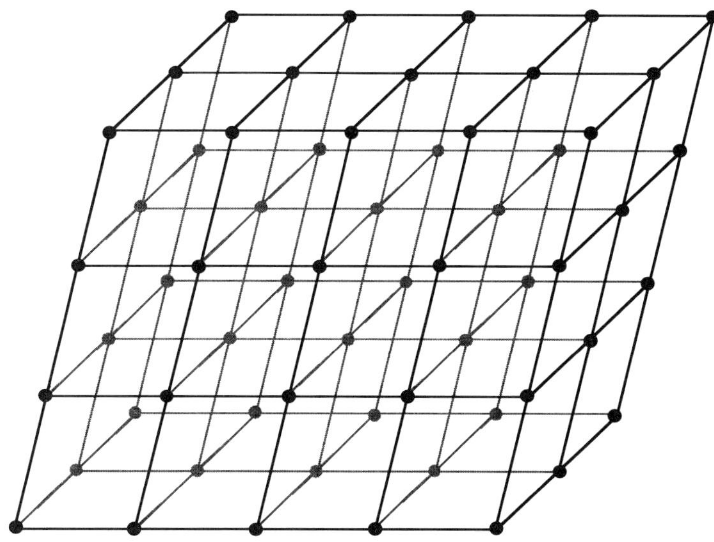

FIG. 15. A point lattice. The light lattice points are those that would not be visible when looking from the front of the lattice. But all the points are equivalent.

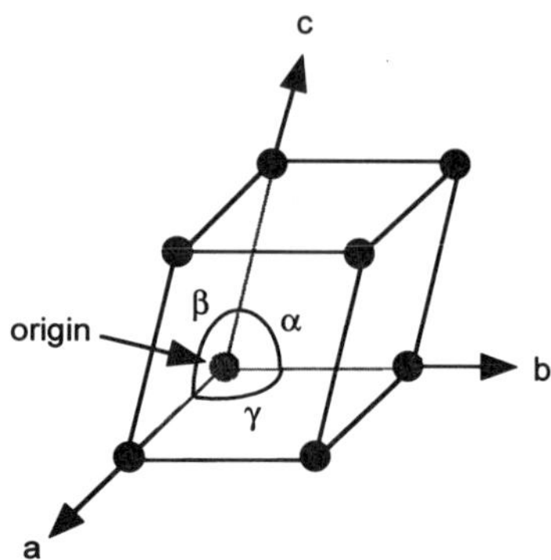

FIG. 16. A unit cell.

the origin. These lengths and angles are called the *lattice parameters* of the unit cell, or sometimes the lattice constants of the cell. But the latter term is not really appropriate because they are not necessarily constants; for example, they can vary with changes in temperature and pressure and with alloying. [Note: We use a, b, and c to indicate the axes of the unit cell; *a*, *b*, and *c* for the lattice parameters, and **a**, **b**, and **c** for the vectors lying along the unit-cell axes.]

2.3. CRYSTAL SYSTEMS AND BRAVAIS LATTICES

You can probably imagine unit cells of many different shapes. However, one of the requirements of a unit cell is that they can be stacked to fill three-dimensional space. Seven unit-cell shapes meet this requirement and are known as the seven *crystal systems*. All crystals can be classified into these seven categories. We have listed the seven crystal systems in Table 3 in order of increasing symmetry. The triclinic cell has the lowest symmetry, and the cubic cell has the highest symmetry. (*Quasicrystals* are not included in this classification. They are a relatively new form of solid matter wherein the atoms are arranged in a three-dimensional pattern that exhibits the traditionally forbidden translational symmetries, e.g., five-fold, seven-fold, etc.)

If we put a lattice point at the corner of each unit cell of the seven crystal systems, we obtain seven different point lattices. However, other arrangements of points also satisfy the requirement of a point lattice; i.e., each point has identical surroundings. Auguste Bravais, in 1848, demonstrated that there are 14 possible point lattices and no more. The 14 *Bravais*

TABLE 3. The Seven Crystal Systems

System	Relationship among Lattice Parameters
Triclinic	$a \neq b \neq c$
	$\alpha \neq \beta \neq \gamma \neq 90°$
Monoclinic	$a \neq b \neq c$
	$\alpha = \gamma = 90°$; $\beta \neq 90°$
Orthorhombic	$a \neq b \neq c$
	$\alpha = \beta = \gamma = 90°$
Tetragonal	$a = b \neq c$
	$\alpha = \beta = \gamma = 90°$
Hexagonal	$a = b \neq c$
	$\alpha = \beta = 90°$; $\gamma = 120°$
Rhombohedral (or trigonal)	$a = b = c$
	$\alpha = \beta = \gamma \neq 90°$
Cubic	$a = b = c$
	$\alpha = \beta = \gamma = 90°$

Lattice Points Per Cell

To readily calculate some of the physical characteristics of crystals, such as atomic density and density of free electrons, we must know the number of lattice points per cell. You may think that such a determination is quite trivial, but it is surprising how many people have difficulty with it! The number of lattice points per cell, N, is given by the equation

$$N = N_i + \frac{N_f}{2} + \frac{N_c}{8} \tag{8}$$

where N_i is the number of lattice points in the interior of the cell (these points "belong" only to one cell), N_f is the number of lattice points on faces (these are shared by two cells), and N_c is the number of lattice points on corners (these are shared by eight cells). For the three cubic unit cells the number of lattice points per cell is

Primitive cubic (cubic P)	1
Body-centered cubic (cubic I)	2
Face-centered cubic (cubic F)	4

All primitive cells have one lattice point per cell. *All* nonprimitive cells have more than one lattice point per cell.

lattices are shown in Fig. 17. These lattices are known interchangeably as Bravais lattices, point lattices, and space lattices.

The lattice symbols given to the Bravais lattices in Fig. 17 have the following meanings:

- P stands for a *primitive* or simple cell, where there is a lattice point at each corner.
- F refers to a *face-centered* cell, where a lattice point is centered on each face, in addition to the corners of the unit cell.
- I is used for *body-centered* cells, where a lattice point is in the center of the cell—in the *interior* of the cell—in addition to the corners of the unit cell.
- A, B, and C refer to *base-centered* cells where lattice points are centered on opposite faces of the cell, in addition to the corners of

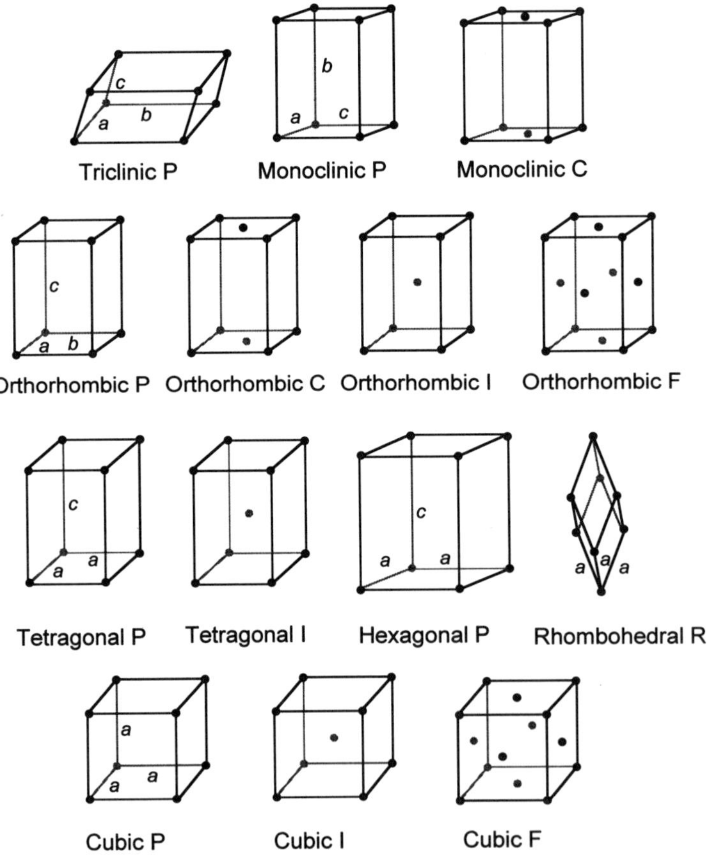

FIG. 17. The 14 Bravais lattices.

the unit cell. (The A face is the face defined by the b and c axes, the B face is defined by the a and c axes, and the C face is defined by the a and b axes. In Fig. 17 only the unit cells having lattice points centered on the C face have been shown.)

- R is used only for the *rhombohedral* system and refers to a primitive cell.

Some texts list only six crystal systems because rhombohedral crystals can always be described in terms of hexagonal axes, so the rhombohedral system is often considered to be a subdivision of the hexagonal system.

So far we have discussed only lattice points. What is the difference between a lattice point and an atom? A lattice point represents equivalent positions in a Bravais lattice. In a real crystal a lattice point may be occupied by one atom or by a group of atoms. In the latter case the atoms are in a fixed relationship with respect to each lattice point. In both cases, the number, composition, and arrangement of atoms is the same for each lattice point. This arrangement is known as the *basis*.

An important difference between lattice points and atoms is that the lattice points tell us nothing about the chemistry or bonding within the crystal; for that we need to include the identity of the atoms and their positions.

2.4. CRYSTAL STRUCTURES

Now we want to consider actual crystals and their structures. The relationship between Bravais lattices and actual crystal structures involves the basis. We can express this relationship as

$$\text{Bravais lattice} + \text{basis} \rightarrow \text{Crystal structure} \tag{9}$$

You can see that the different crystal structures are built on the framework of one of the 14 Bravais lattices and contain a basis consisting of a number of atoms. In the following sections we group crystal structures in terms of their basis. This approach may be somewhat different from that which is used to describe structures in introductory materials science classes, but it will help you when we describe the structure factor in Sec. 2.8.

2.4.1. One Atom per Lattice Point

The simplest crystal structures are those in which the basis is one atom located on each lattice point of a Bravais lattice. (Remember, each atom must be of the same kind.) Let's consider the crystal structures based on the three cubic Bravais lattices in which each has a basis consisting of one atom. Using Eq. (9) we obtain for the primitive cubic Bravais lattice:

Primitive cubic (cubic P) lattice + one atom → Simple cubic (sc) structure

The sc structure is illustrated in Fig. 18. There is one atom per cell in the sc structure and this atom is located at the origin; i.e., its coordinates are 0,0,0. (Our choice of the origin of the unit cell is entirely arbitrary, as you will see.) The simple cubic structure is uncommon; no important metals have this structure. α-Polonium (Po) is the only element that crystallizes in the simple cubic structure, although some nonequilibrium phases obtained by rapid solidification or mechanical alloying also exhibit this structure.

Now let's consider the body-centered cubic Bravais lattice with a basis consisting of one atom. Using Eq. (9) we obtain

Body-centered cubic (cubic I) lattice + one atom →
Body-centered cubic (bcc) structure

The bcc crystal structure is shown in Fig. 19. There are two atoms per cell, located at 0,0,0 and $\frac{1}{2},\frac{1}{2},\frac{1}{2}$. The two atom positions are related by the body-centering translation $\frac{1}{2},\frac{1}{2},\frac{1}{2}$ (i.e., a translation of half a lattice parameter along the a axis, half a lattice parameter along the b axis, and half a lattice parameter along the c axis). If the body-centering translation is

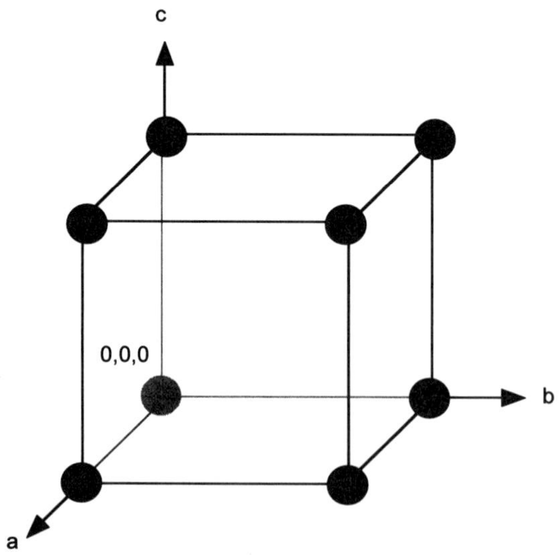

FIG. 18. Simple cubic structure.

Coordinates of Points

The location of certain points, such as the position of lattice points or atoms, in the unit cell is straightforward. It is usual to choose a right-handed coordinate system, (Fig. 16), where the origin is located at the back left-hand corner of the cell as you look at it. The distance is measured in terms of how many lattice parameters we must move along the a, b, and c axes to get from the origin to the point we are interested in. The coordinates are written as the three distances, with commas separating the numbers.

applied to the atom at 0,0,0, the body-centered atom is reproduced. If the body-centering translation is applied to the atom at $\frac{1}{2},\frac{1}{2},\frac{1}{2}$, then one of the corner atoms is reproduced. Several metals exist in the bcc structure, including sodium (Na), chromium (Cr), α-iron (α-Fe), molybdenum (Mo), and tungsten (W).

The face-centered cubic (fcc) structure in Fig. 20 is based on the face-centered cubic (cubic F) Bravais lattice with a basis of one atom:

Face-centered cubic (cubic F) lattice + one atom →
Face-centered cubic (fcc) structure

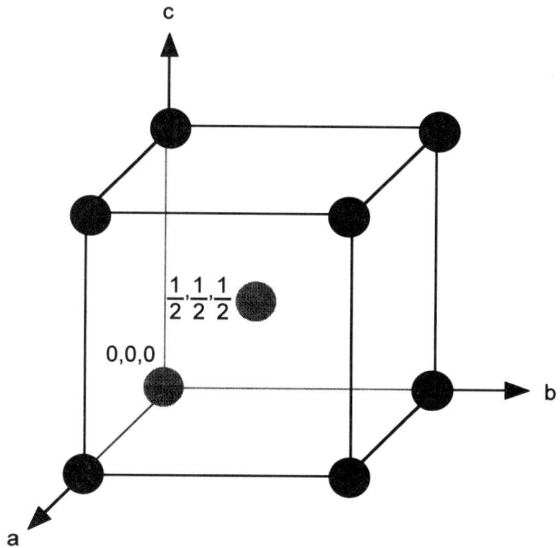

FIG. 19. Body-centered cubic structure.

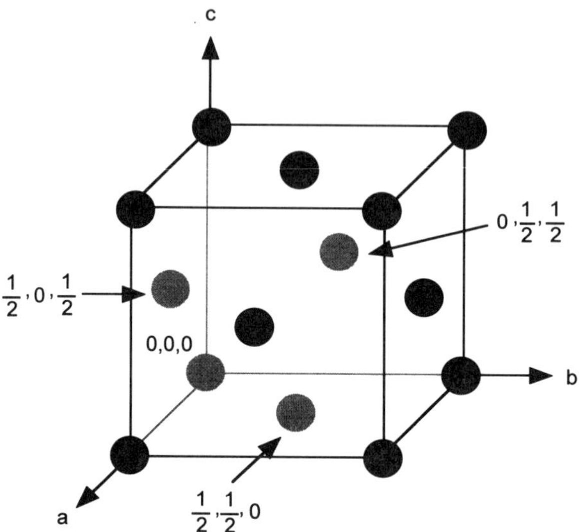

FIG. 20. Face-centered cubic structure.

The fcc structure has four atoms per cell, located at 0,0,0; $\frac{1}{2}$,$\frac{1}{2}$,0; $\frac{1}{2}$,0,$\frac{1}{2}$; and 0,$\frac{1}{2}$,$\frac{1}{2}$. If the face-centering translations $\frac{1}{2}$,$\frac{1}{2}$,0; $\frac{1}{2}$,0,$\frac{1}{2}$; and 0,$\frac{1}{2}$,$\frac{1}{2}$ are repeatedly applied to any atom in the cell, the positions of all the atoms in the crystal structure will be reproduced. The fcc structure is exhibited by several metals including calcium (Ca), copper (Cu), gold (Au), nickel (Ni), and silver (Ag). [Note: When referring to the Bravais lattice we write, for example, face-centered cubic, but when referring to a crystal with the face-centered cubic structure we write fcc.]

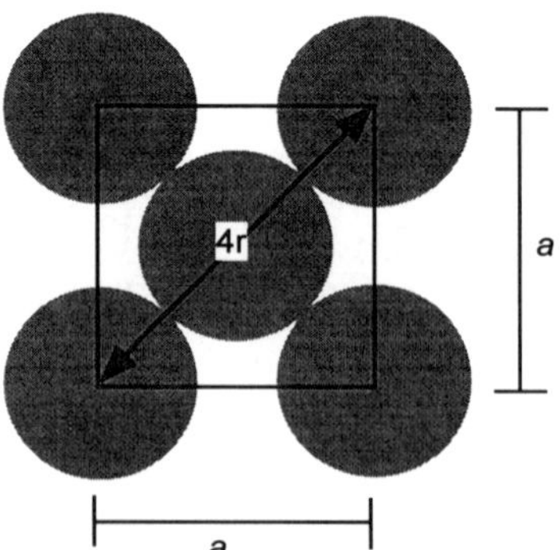

FIG. 21. Atoms touch across the face diagonal in the fcc structure.

Close-Packed Structures

A close-packed structure is one that has the maximum volume of the unit cell occupied by atoms. The fraction of the unit cell occupied by atoms can be determined by calculating the atomic packing factor (APF) from the following equation:

$$\text{APF} = \frac{\text{number of atoms per cell} \times \text{volume of one atom}}{\text{volume of unit cell}} \tag{10}$$

We will go through the calculation of the APF for the fcc structure; you may calculate for the other cubic structures as an exercise.

Look back at the fcc structure in Fig. 20:

- The number of atoms per cell is 4.
- The volume of each atom is $(4/3)\pi r^3$ (we assume in these calculations that the atoms are rigid spheres with a radius r).
- The unit-cell volume is a^3.

We can rewrite the lattice parameter a in terms of r (Fig. 21), where the figure shows that the atoms touch across the face diagonal of the cell. Therefore, $\sqrt{2}a = 4r$, or $a = 2\sqrt{2}r$. Using Eq. (10) we obtain

$$\text{APF} = \frac{4(\frac{4}{3}\pi r^3)}{(2\sqrt{2}r)^3} = 0.74$$

For the fcc structure, the atomic packing factor is 0.74. This is the *maximum* possible value for packing spheres of the same size. Crystal structures with an APF of 0.74 are called close-packed structures. Of the three cubic structures only the fcc structure is close-packed.

2.4.2. Two Atoms of the Same Kind per Lattice Point

The Hexagonal Close-Packed Structure

Many metals have the hexagonal close-packed (hcp) structure, including magnesium (Mg), titanium (Ti), zinc (Zn), and cadmium (Cd). The hcp structure is built on the hexagonal Bravais lattice with a basis consisting of two identical atoms associated with each lattice point. Using the relationship in Eq. (9), we have

Hexagonal P lattice + two atoms → Hexagonal close-packed (hcp) structure

Figure 22a shows the arrangement of atoms in the hcp structure. There are two atoms per unit cell, located at 0,0,0 and $\frac{2}{3},\frac{1}{3},\frac{1}{2}$. However, in the

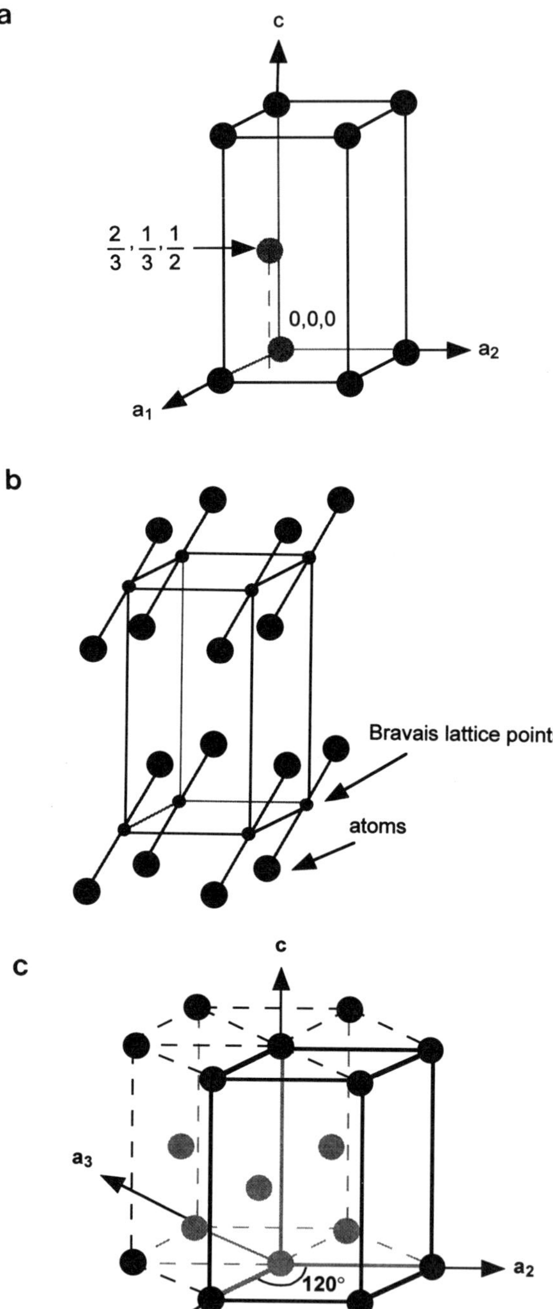

FIG. 22. Hexagonal close-packed structure.

hexagonal Bravais lattice there is only one lattice point per cell (because the cell is primitive). Figure 22b shows the hcp structure redrawn (of course, all the atoms are still in same place) with the origin of the unit cell shifted upward so that the point 1,0,0 in the new cell is now midway between the atoms at 1,0,0 and $\frac{2}{3},\frac{1}{3},\frac{1}{2}$ in the old cell. Now, if a Bravais lattice point is located on each corner of the new cell, then a pair of atoms is associated with each Bravais lattice point. To get a better idea of the hexagonal symmetry of the hcp structure, look at Fig. 22c. The hcp structure, as the name suggests, is a close-packed structure (APF = 0.74). Only the hcp and fcc structures are close-packed.

The Diamond Cubic Structure

Another important structure with two atoms per lattice point is the diamond cubic. This structure is adopted by diamond (one of the crystalline forms of carbon) and the industrially important elemental semiconductors silicon (Si) and germanium (Ge).

The diamond cubic structure (Fig. 23) is based on the face-centered cubic Bravais lattice with a basis of two atoms:

Face-centered cubic (cubic F) lattice + two atoms $\rightarrow$ Diamond cubic structure

There are eight atoms per cell in the diamond cubic structure, located at 0,0,0; $\frac{1}{2},\frac{1}{2}$,0; $\frac{1}{2}$,0,$\frac{1}{2}$; 0,$\frac{1}{2},\frac{1}{2}$; $\frac{1}{4},\frac{1}{4},\frac{1}{4}$; $\frac{3}{4},\frac{3}{4},\frac{1}{4}$; $\frac{3}{4},\frac{1}{4},\frac{3}{4}$; $\frac{1}{4},\frac{3}{4},\frac{3}{4}$, or, in terms of the face-centering

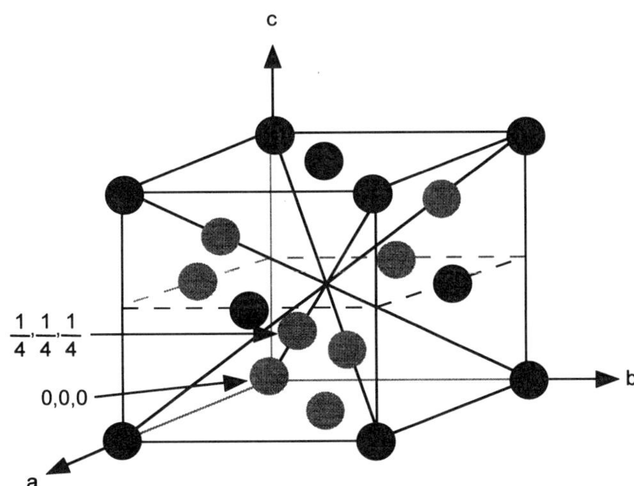

FIG. 23. Diamond cubic structure.

translations, 0,0,0 + face-centering translations, and $\frac{1}{4},\frac{1}{4},\frac{1}{4}$ + face-centering translations.

Although the atoms at these two sets of positions are chemically identical (they are all carbon atoms in diamond or all silicon atoms in a crystal of silicon) the lattice positions are not equivalent. If the two lattice positions were equivalent, it would be possible to move from one to another by the same translation vector. The vector $\mathbf{v}_1$ in Fig. 24 would move the atom located at 0,0,0 to the point $\frac{1}{2},\frac{1}{2}$,0, which is also a lattice position in the diamond cubic structure. The vector $\mathbf{v}_2$ in Fig. 24 would move an atom at 0,0,0 to the point $\frac{1}{4},\frac{1}{4},\frac{1}{4}$. However, if the point $\frac{1}{4},\frac{1}{4},\frac{1}{4}$ is operated on by $\mathbf{v}_2$ it would translate it to the coordinates $\frac{1}{2},\frac{1}{2},\frac{1}{2}$ (a position in the body center of the diamond cubic unit cell). The point $\frac{1}{2},\frac{1}{2},\frac{1}{2}$ is not a lattice position in the diamond cubic structure.

If we consider only the atom at 0,0,0 and use the face-centering translations, we produce an fcc crystal structure. If we now consider only the atom at $\frac{1}{4},\frac{1}{4},\frac{1}{4}$ and perform the face-centering translations, we also produce an fcc crystal structure. So we can think of the diamond cubic

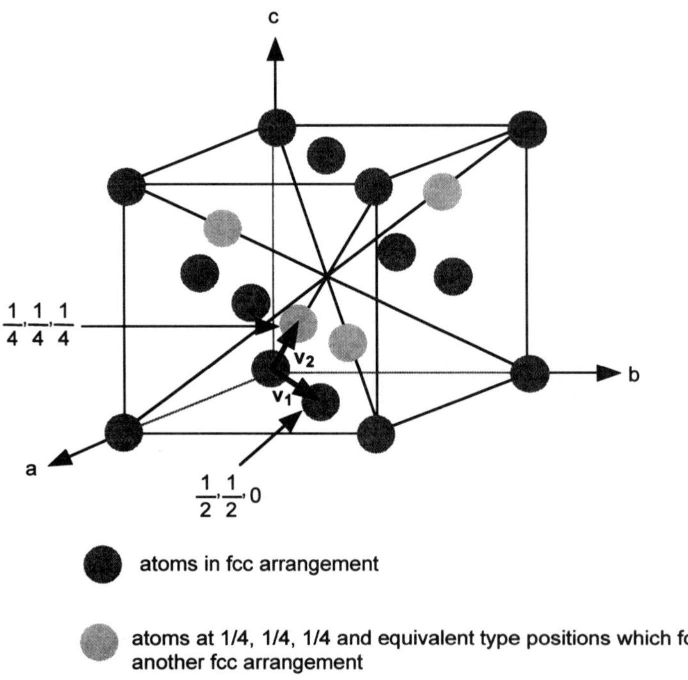

FIG. 24. Illustration of the diamond cubic structure showing the atoms which form two face-centered cubic arrangements.

Names of Crystal Structures

As we go through this chapter, and when you take other materials science classes, you will find that many crystal structures are named after a particular material (often a naturally occurring mineral) that exhibits the structure. There are no systematic names for crystal structures. In chemistry, for example, organic compounds are named with a system recommended by the International Union of Pure and Applied Chemistry (IUPAC). The IUPAC system is a systematic way of naming many organic compounds on sight, and the name indicates the structure of the compound. A similar system is not used for naming crystal structures. However, in the next section we introduce a systematic notation for crystal structures, which can be very useful. Table 4 lists the names of some common crystal structures. For example, MgO (which has a melting point of 2800°C) has the same structure as NaCl (melting point of 801°C). Even though two or more materials may share the same crystal structure they may have different properties. The properties of a material depend on the types of atoms present, the way the atoms are held together (bonding), the crystal structure, and the defects present. You can also see from the table that ZnS exists in both the zinc blende (sphalerite) and wurtzite structures. This phenomenon is known as *polymorphism* and is exhibited by many materials.

structure as consisting of two *interpenetrating* face-centered cubic lattices, the origin of one displaced by $\frac{1}{4},\frac{1}{4},\frac{1}{4}$ from the other.

It is important to realize that, although the fcc structure and the diamond cubic structure both have the same Bravais lattice, they are very different structures and, consequently, materials with a diamond cubic structure have properties significantly different from those with an fcc

TABLE 4. Names of Some Common Crystal Structures

Name of Structure	Corresponding Mineral	Other Materials with the Same Structure
Rocksalt	NaCl	KCl, LiF, MgO, NiO, CaO, TiN
Zinc blende (sphalerite)	ZnS	BeO, β-SiC, GaAs
Rutile	TiO_2	GeO_2, SnO_2
Corundum	Al_2O_3	Fe_2O_3, Cr_2O_3, Ti_2O_3
Wurtzite	ZnS	ZnO, α-SiC, AlN
Perovskite	$CaTiO_3$	$BaTiO_3$, $SrTiO_3$
Spinel	$MgAl_2O_4$	$FeAl_2O_4$, $ZnAl_2O_4$

structure. One significant difference between the two structures is the atomic packing factor. The APF is only 0.34 for the diamond cubic structure: remember it was 0.74 for the fcc structure. You will also see that the structure factors (selection rules for the observance of reflections in the x-ray diffraction pattern) are different for the two structures.

2.4.3. Two Different Atoms per Lattice Point

We describe four crystal structures in this section, known by the common names

- Cesium chloride (CsCl)
- Sodium chloride (NaCl, or rocksalt)
- Zinc blende (ZnS, or sphalerite)
- Wurtzite (ZnS)

Cesium Chloride Structure

In addition to CsCl, many other compounds have the "CsCl structure," including CsBr, NiAl, the ordered form of β-brass (CuZn), and CuPd. The CsCl structure is shown in Fig. 25. The unit cell contains two atoms. For

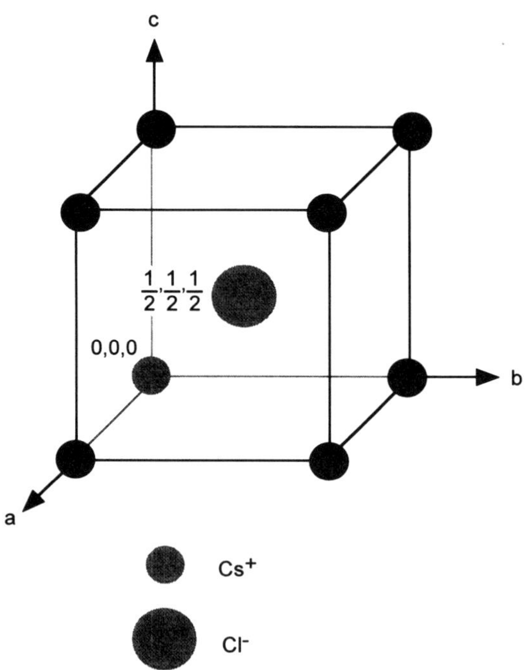

FIG. 25. Cesium chloride structure.

CsCl, which is strongly ionic, it would be better to refer to ions rather than atoms, but remember that compounds with nonionic bonding can also have this structure. In the CsCl structure there is a cesium ion at 0,0,0 and a chlorine ion at $\frac{1}{2},\frac{1}{2},\frac{1}{2}$ forming the basis. The Bravais lattice is primitive cubic (cubic P):

Primitive cubic (cubic P) lattice + two atoms → CsCl structure

A common mistake when assigning the Bravais lattice to the CsCl structure is to say that the Bravais lattice is body-centered cubic. However, in the bcc crystal structure, which we saw in Sec. 2.4.1 is built on the body-centered cubic Bravais lattice, all the atoms must be of the *same* type and the body-centering translation must join identical atoms. In CsCl if we apply the body-centering translation to the Cs^+ ion at 0,0,0 we would get to a position occupied by a Cl^- ion. So in CsCl, where there are two kinds of atoms, the Bravais lattice cannot be body-centered cubic.

We can describe the CsCl structure as two interpenetrating simple cubic structures—one of Cs^+ ions and one of Cl^- ions, one displaced by $\frac{1}{2},\frac{1}{2},\frac{1}{2}$ with respect to the other.

Sodium Chloride Structure

Many materials exist in the sodium chloride (NaCl) or rocksalt crystal structure. In addition to many of the halides, including AgCl, KBr, LiF, NaI, many ceramics, including MgO, CaO, NiO, TiN, ZrN, TiC, ZrC, and ZrB, have the NaCl structure. The NaCl structure (Fig. 26) is based on the face-centered cubic Bravais lattice with a basis of two atoms at 0,0,0 and $\frac{1}{2}$,0,0.

Face-centered cubic (cubic F) lattice + two atoms → NaCl structure

There are a total of eight atoms per cell, four Na and four Cl. (Strictly speaking, for the actual compound NaCl we should refer to ions, Na^+ and Cl^-, not atoms, because the compound is completely ionized.) The Na^+ ions are located at positions 0,0,0; $\frac{1}{2},\frac{1}{2}$,0; $\frac{1}{2}$,0,$\frac{1}{2}$; 0,$\frac{1}{2},\frac{1}{2}$; and the Cl^- ions are located at $\frac{1}{2}$,0,0; 0,$\frac{1}{2}$,0; 0,0,$\frac{1}{2}$; $\frac{1}{2},\frac{1}{2},\frac{1}{2}$.

The sodium ions are in a face-centered cubic arrangement and if we apply the face-centering translations to the chlorine ion at $\frac{1}{2}$,0,0 we can see that the chlorine ions are also in a face-centered cubic arrangement. So we can describe the NaCl structure as two interpenetrating face-centered cubic lattices, the origin of one displaced from the other by $\frac{1}{2}$,0,0.

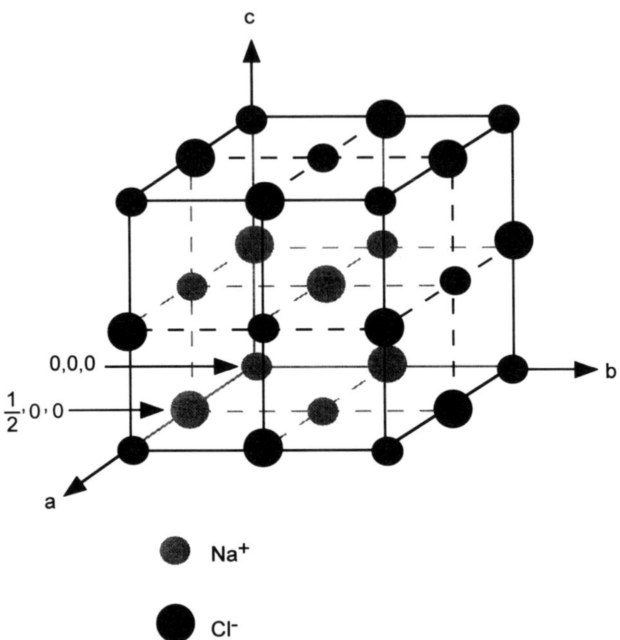

FIG. 26. Sodium chloride structure.

Zinc Blende Structure

The zinc blende, or sphalerite, structure is shown in Fig. 27. In this structure the arrangement of atoms is the same as in the diamond cubic structure in Fig. 23. But the zinc blende structure has two types of atoms in the unit cell (classically Zn and S). The structure is built on the face-centered cubic Bravais lattice with two atoms per lattice point, one at 0,0,0 and the other at $\frac{1}{4},\frac{1}{4},\frac{1}{4}$.

Face-centered cubic (cubic F) lattice + two atoms → Zinc blende structure

There are a total of eight atoms per cell, four Zn and four S, and these are located at

S atoms at	0,0,0	$\frac{1}{2},\frac{1}{2},0$	$\frac{1}{2},0,\frac{1}{2}$	$0,\frac{1}{2},\frac{1}{2}$
Zn atoms at	$\frac{1}{4},\frac{1}{4},\frac{1}{4}$	$\frac{3}{4},\frac{3}{4},\frac{1}{4}$	$\frac{3}{4},\frac{1}{4},\frac{3}{4}$	$\frac{1}{4},\frac{3}{4},\frac{3}{4}$

or, in terms of the face-centering translations,

S	0,0,0 + face-centering translations
Zn	$\frac{1}{4},\frac{1}{4},\frac{1}{4}$ + face-centering translation

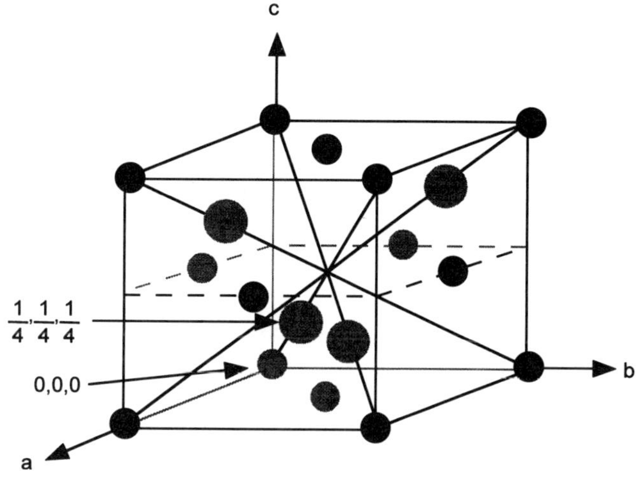

FIG. 27. Zinc blende structure.

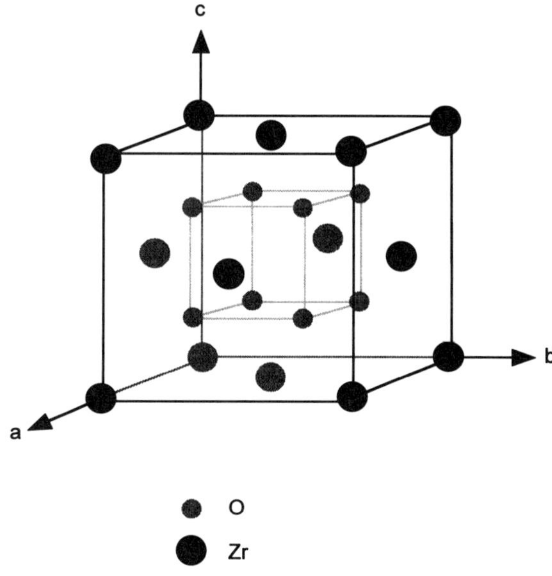

FIG. 28. Fluorite structure.

2.4.4. More than Two Atoms per Lattice Point

In the experimental modules in Part II we have concentrated on crystal structures that have either one or two atoms per lattice point. But you should be aware that there are crystal structures with more than two atoms associated with each lattice point. Of course, the structures are still based on 1 of the 14 Bravais lattices shown in Fig. 17. In this section we present just three examples of materials having crystal structures more complex than those already considered.

The first example is cubic zirconium oxide (ZrO_2), or cubic zirconia as it is commonly known. Cubic zirconia is a hard ceramic that exists in what

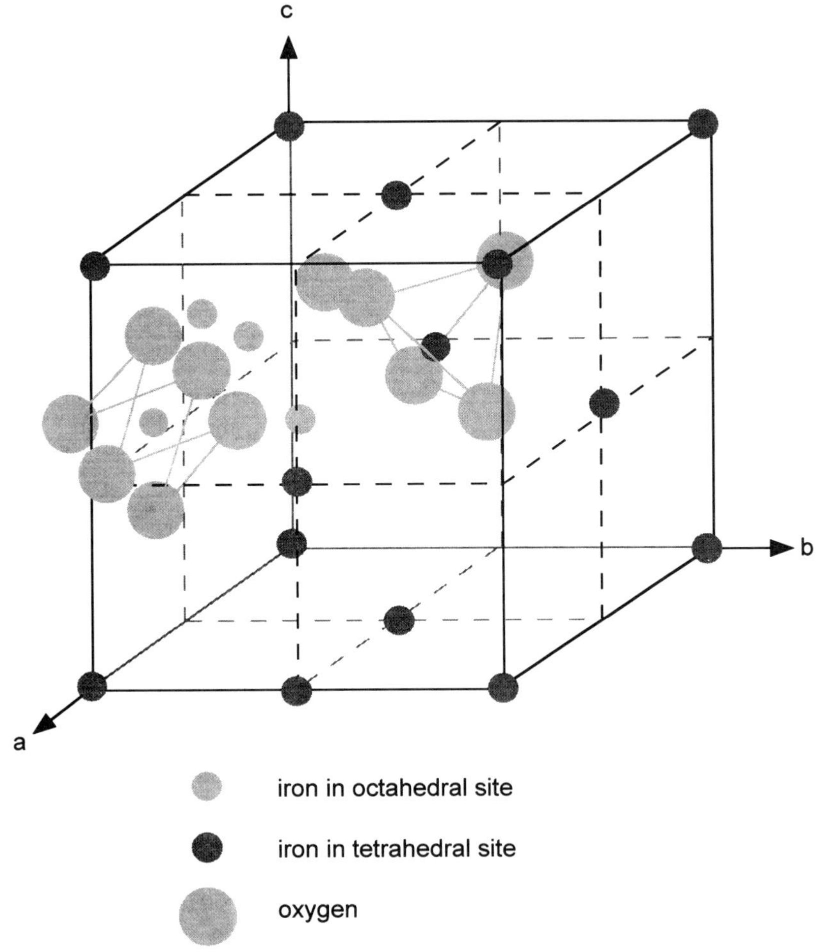

FIG. 29. Part of the spinel structure. There are 32 octahedral sites and 64 tetrahedral sites per unit cell.

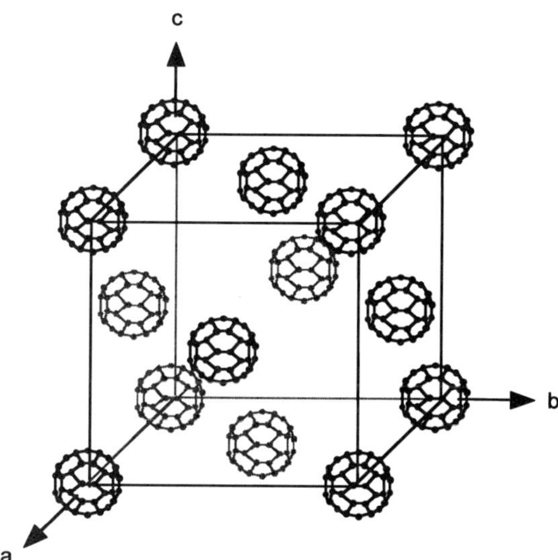

FIG. 30. Structure of solid C_{60}.

is called the fluorite structure (Fig. 28). The structure is built on the face-centered cubic Bravais lattice with a basis of three atoms (two oxygen and one zirconium).

The second example is the spinel structure. The mineral spinel is $MgAl_2O_4$, but many other materials exist in its structure. One example is the magnetic oxide Fe_3O_4 (known by the common name magnetite). The structure of magnetite is shown in Fig. 29. The Bravais lattice is face-centered cubic once again, but this time there are 14 atoms associated with each lattice point. (There are 56 atoms per unit cell.)

The last example is the structure of solid C_{60}. The C_{60} molecule is quite familiar to many people, even though it was only discovered in 1985, because its shape is characteristic of a geodesic dome and is called a "buckyball." In the solid state, the C_{60} molecules form a crystalline structure in which they are arranged in a face-centered cubic array as shown in Fig. 30. At each lattice point there is one C_{60} molecule, so there are 60 atoms associated with each lattice point, or 240 atoms per unit cell!

2.5. NOTATION FOR CRYSTAL STRUCTURES

As noted, there is no systematic sequence for naming crystal structures. However, a *systematic notation* for describing crystal structures has been developed by W. B. Pearson (*The Crystal Chemistry and Physics of Metals*

TABLE 5. Symbols for the 14 Bravais Lattices

Symbol	System	Lattice Symbol
aP	Triclinic (anorthic)	P
mP	Simple monoclinic	P
mC	Base-centered monoclinic	C
oP	Simple orthorhombic	P
oC	Base-centered orthorhombic	C
oF	Face-centered orthorhombic	F
oI	Body-centered orthorhombic	I
tP	Simple tetragonal	P
tI	Body-centered tetragonal	I
hP	Hexagonal	P
hR	Rhombohedral	R
cP	Simple cubic	P
cF	Face-centered cubic	F
cI	Body-centered cubic	I

and Alloys, Wiley-Interscience, New York, 1972). This system classifies structures according to crystal system, Bravais lattice, and number of atoms per unit cell. The symbols give successively the crystal system, the Bravais lattice symbol, and the number of atoms per unit cell according to the designation in Table 5. For the crystal structures in Sec. 2.4 the Pearson symbols are given in Table 6.

Even though you can determine the crystal system, the Bravais lattice, and the number of atoms from this notation, you will not be able to differentiate different structures with the same notation. For example, cF8

TABLE 6. Pearson Symbols for Some Crystal Structures

Crystal Structure	Pearson Symbol
Simple cubic	cP1
Body-centered cubic	cI2
Face-centered cubic	cF4
Hexagonal close-packed	hP2
Cesium chloride	cP2
Sodium chloride	cF8
Diamond cubic	cF8
Zinc Blende	cF8
Fluorite	cF12
Spinel	cF56
Solid C_{60}	cF240

can refer to the structures of sodium chloride, diamond cubic, or zinc blende, which differ from one another. However, this appears to be the best possible notation for the moment and is extensively used by the International Centre for Diffraction Data when referring to crystal structures, as you will see in Experimental Module 8.

2.6. MILLER INDICES

It is essential to have a way of uniquely describing different crystal planes in a lattice. Miller indices provide us with such a way. They are determined as follows:

1. Identify the points at which the plane intersects the a, b, and c axes. The intercept is measured in terms of fractions or multiples of the lattice parameters. If the plane moves through the origin, then the origin must be moved. We will show you how this works a little later.
2. Take reciprocals of the intercepts. This is done because if the plane never intersects one of the axes the intercept will be at ∞. We don't want to have ∞ in the indices so we take reciprocals.
3. Clear the fractions. We only want whole numbers in the indices, but don't reduce the numbers to the lowest integers.
4. Enclose the resulting numbers in parentheses ( ). Negative numbers are written with a bar over the number ($\bar{1}$), and pronounced, for example, as bar one.

These steps give us the Miller indices of any plane in any crystal system, whether cubic, orthorhombic, or triclinic. Also note that the interaxial angles are not important in this case—the A face has indices (100) in all crystal systems, and the C face has indices (001) in all crystal systems. The actual values of the lattice parameters *a, b,* and *c* are also unimportant.

Let's determine the Miller indices of the planes A and B in Fig. 31.

For plane A:

	a	b	c
Intercepts	1	1	1
Reciprocals	1	1	1
Clear fractions (not necessary in this case)	1	1	1

Therefore, plane A has Miller indices (111).

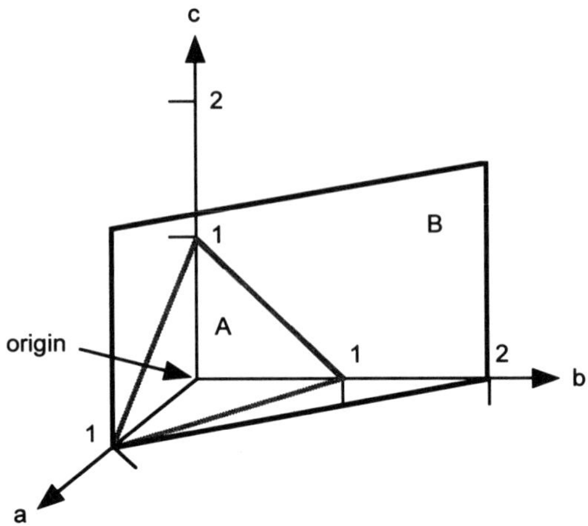

FIG. 31. Determination of Miller indices of planes.

For plane B:

	a	b	c
Intercepts	1	2	∞
Reciprocals	1	$\frac{1}{2}$	0
Clear fractions	2	1	0

Therefore, the Miller indices for plane B are (210).

Let's now look at what happens if the plane goes through the origin, as does plane C in Fig. 32. To index this plane we need to move the origin. Remember that the origin of the unit cell is arbitrarily chosen and can be taken to be any point. So let's move the origin to the point that originally had coordinates 0,1,0. Now we follow the same procedure as before, only using our new origin.

For plane C:

	a	b	c
Intercepts	∞	−1	∞
Reciprocals	0	−1	0
Clear fractions	0	−1	0

So plane C is the $(0\bar{1}0)$ plane.

If, however, you want to compare the Miller indices of planes in a unit cell, the origin must be the same for each plane. In such a situation, if the

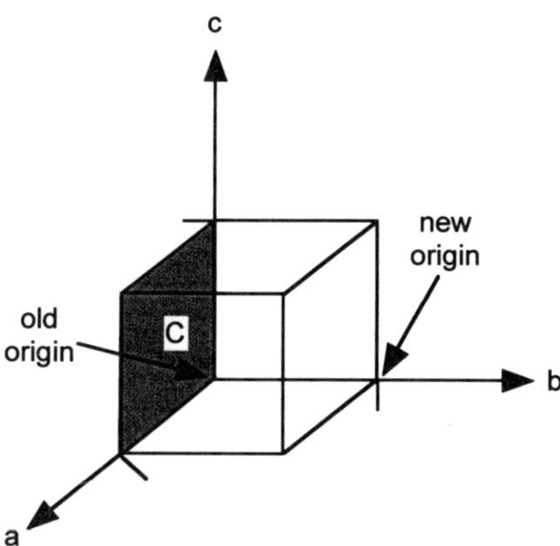

FIG. 32. Determination of Miller indices of a plane when it passes through the origin.

plane goes through the origin, instead of shifting the origin you can mentally shift the position of the plane to be parallel to the plane under consideration.

Look again at the cubic unit cell in Fig. 32. Notice that plane C is a face of the cube. If you closed your eyes and we rotated the cube 90° clockwise, when you opened your eyes you would not know that the cube had been moved. Hence, the face planes of a cube are all symmetrically related. There are six different face planes, but they are all equivalent. When a set of planes is equivalent, they are known as a *family* of planes. The faces of the cube form the family of {100} planes. Note that now we use a different

Multiplicity

As we have just seen, there are six planes in the {100} family of planes for a cubic crystal. These planes all have the same spacing. We can define the *multiplicity factor,* which is the number of planes in a family that have the same spacing. The multiplicity or multiplicity factor for the {100} family of planes of a cubic crystal is 6. The multiplicity factors are different for different planes, and these values for the cubic and hexagonal systems are listed in Appendix 6.

The multiplicity is important because in a polycrystalline or powder specimen consisting of many randomly oriented grains diffraction can occur, for example, from all the {100} planes. The intensity of a reflection therefore depends on the multiplicity. You will need to use the multiplicity factor in Experimental Module 7.

type of bracket to indicate a family of planes. Parentheses ( ) are used for a specific plane, and braces { } are used for a family of planes. The six planes in the {100} family of planes for a cubic unit cell are (100), (010), (001), $(\bar{1}00)$, $(0\bar{1}0)$, and $(00\bar{1})$. For a tetragonal unit cell the {100} family of planes consists of four planes that have Miller indices (100), (010), $(\bar{1}00)$, and $(0\bar{1}0)$, and the {001} family of planes consists of planes (001) and $(00\bar{1})$. In the tetragonal system the {100} planes are not equivalent to the {001} planes since $a \neq c$.

When you start to work through the experimental modules you will find that there are peaks in the x-ray diffraction pattern that correspond to planes of the form ($nhnknl$), where n is an integer. What is the relationship between the plane (hkl) and the planes ($nhnknl$)? The two sets of planes are parallel, but the spacing of the ($nh\ nk\ nl$) planes is $1/n$ the spacing of the (hkl) plane. As shown in Fig. 33, the (100) and (200) planes are parallel, but the spacing of the (200) planes is half the spacing of the (100) planes. For example, in magnesium oxide (MgO) the spacing of the (200) planes is 0.2106 nm; the spacing of the (400) planes is 0.1053 nm.

The distance, d, between adjacent planes in the set (hkl) for a crystal in the cubic system is

$$\frac{1}{d^2} = \frac{h^2 + k^2 + l^2}{a^2} \tag{11}$$

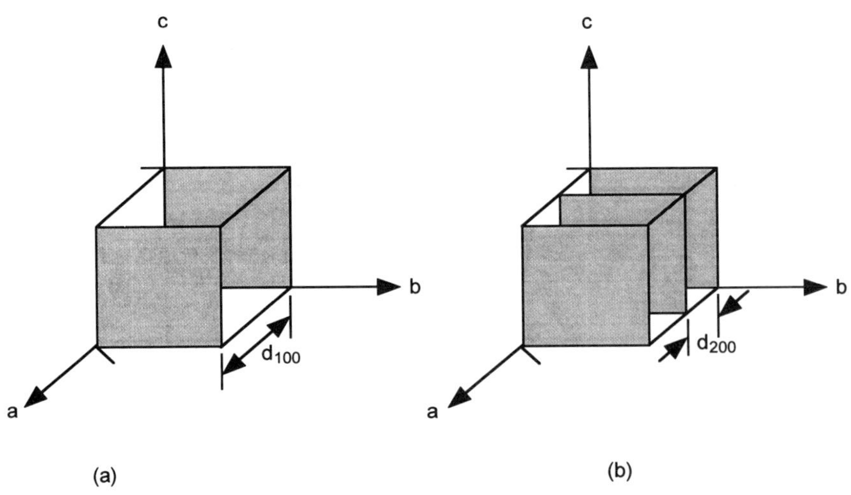

FIG. 33. Difference between the (100) and (200) planes.

This relationship, which can be derived quite simply by using analytical geometry, is very useful, and you will use it in Experimental Module 1 and others. Equation (11) applies only to cubic systems, but similar equations can be formulated for the other crystal systems and they are listed in Appendix 1.

The density of atoms on a particular plane (known as the planar density) can be determined from the equation

$$\text{Planar density} = \frac{\text{number of atoms on the plane}}{\text{area of the plane}} \quad (12)$$

For example, the planar density on the (100) plane of a material with the fcc structure, shown in Fig. 34, is

$$\text{Planar density} = \frac{2}{a^2} = \frac{2}{(2\sqrt{2}\,r)^2} = \frac{1}{4r^2}$$

Copper has an atomic radius of 0.128 nm, and the density of atoms on its (100) plane is 15.26 atoms/nm^2. Planes of the lowest indices (called low-index planes) have the largest *d* spacing and the greatest density of lattice points (and atoms), and, as we shall see, they produce the most intense reflections in the x-ray diffraction patterns.

A slightly different system is used for indexing planes in the hexagonal system. Although most of the important characteristics of hcp crystals can

Notation for Directions in a Crystal

Directions in a crystal can be determined in a very straightforward way if the coordinates *u,v,w* of two points lying along that direction are known. If a line connects two points *A* and *B*, then the direction of that line is $(u,v,w)_B - (u,v,w)_A$. For example, if a line points from the origin 0,0,0 (point *A*) to a point *B* located at 1,1,1, then the direction of that line is [111]. Note that for a specific direction we use brackets [ ] to enclose the indices (no commas between them). Keeping the same origin, the line $[\bar{1}\,\bar{1}\,\bar{1}]$ is antiparallel to the line [111]. In the same way that we can have a family of planes, we can also have a family of directions. The [111] direction is across the body diagonal of the cube. A cube has four body diagonals, and if we count the antiparallel directions there are a total of eight body diagonal directions. The indices of these directions are [111], $[\bar{1}\bar{1}\bar{1}]$, $[11\bar{1}]$, $[1\bar{1}1]$, $[\bar{1}11]$, $[\bar{1}\bar{1}1]$, $[\bar{1}1\bar{1}]$, and $[1\bar{1}\bar{1}]$. This family of directions is the <111> family. Note that we use angled brackets < > for a family of directions. In a cubic crystal the [*hkl*] direction is perpendicular to the (*hkl*) plane.

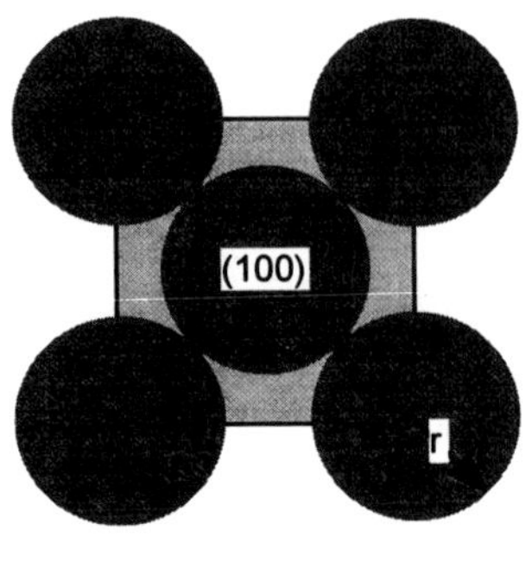

Planar density = $\frac{1}{4r^2}$

FIG. 34. Atom arrangement on the (100) plane in an fcc crystal structure.

be described without the aid of this other system, we feel that you need to be familiar with it. (You will find it particularly useful if you do any electron diffraction of hexagonal materials in the transmission electron microscope.)

The hexagonal unit cell is described by three vectors, $\mathbf{a}_1$, $\mathbf{a}_2$, and $\mathbf{c}$. Vectors $\mathbf{a}_1$ and $\mathbf{a}_2$ lie at 120° to one another in the same plane, called the basal plane, and $\mathbf{c}$ is perpendicular to the basal plane, as shown in Fig. 22c. To show the symmetry of the hexagonal lattice, we often extend the unit cell, as indicated by the dashed lines in Fig. 22c. A third vector, $\mathbf{a}_3$, which also lies in the basal plane at 120° to $\mathbf{a}_1$ or $\mathbf{a}_2$, can be drawn in this extended cell. The indices of a plane in the hexagonal system, written (*hkil*), refer to these four axes and are known as the *Miller–Bravais* indices of the plane. To determine the Miller–Bravais indices of a plane in a hexagonal unit cell, follow exactly the same procedure for determining Miller indices in the cubic system. Figure 35 shows the Miller–Bravais indices of some of the low-index planes and directions in the hexagonal system.

Since the intercepts of a plane on the a_1 and a_2 axes determine its intercept on the a_3 axis, the value of i depends on the values of both h and k. The relationship between these three indices is

$$h + k = -i \tag{13}$$

Thus, it is easy to convert between the four-index Miller–Bravais system and the three-index Miller system. However, the advantage of the four-index system—and the main reason for its use—is that similar planes have similar indices (as we saw for the Miller system for a cubic

Notation for Directions in a Hexagonal Crystal

The directions in a hexagonal cell can be determined in exactly the same way as for a cubic unit cell. We use a three-index notation in which the indices refer to vectors $\mathbf{a}_1$, $\mathbf{a}_2$, and $\mathbf{c}$. As shown in Fig. 35, the directions in the hexagonal system can also be represented by a four-index notation, based on the vectors $\mathbf{a}_1$, $\mathbf{a}_2$, $\mathbf{a}_3$, and $\mathbf{c}$. The three-index notation is often preferred, but the four-index notation has the advantage that similar directions have similar indices. The two notations are related as follows. If $[UVW]$ are the indices of a direction referred to three axes and $[uvtw]$ are the four-axis indices, then

$U = u - t$	$u = (2U - V)/3$
$V = v - t$	$v = (2V - U)/3$
$W = w$	$t = -(u + v) = -(U + V)/3$
	$w = W$

So the direction [100] in the three-index notation becomes $[2\bar{1}\bar{1}0]$ in the four-index notation.

lattice). For example, the planes $(10\bar{1}0)$, $(01\bar{1}0)$, $(\bar{1}100)$, $(\bar{1}010)$, $(0\bar{1}10)$, $(1\bar{1}00)$ are the six side planes (called prism planes) of the hexagonal lattice; these are of similar type. In the Miller system, however, they are written (100), (010), $(\bar{1}10)$, $(\bar{1}00)$, $(0\bar{1}0)$, $(1\bar{1}0)$ and they definitely do not appear to be of a similar type.

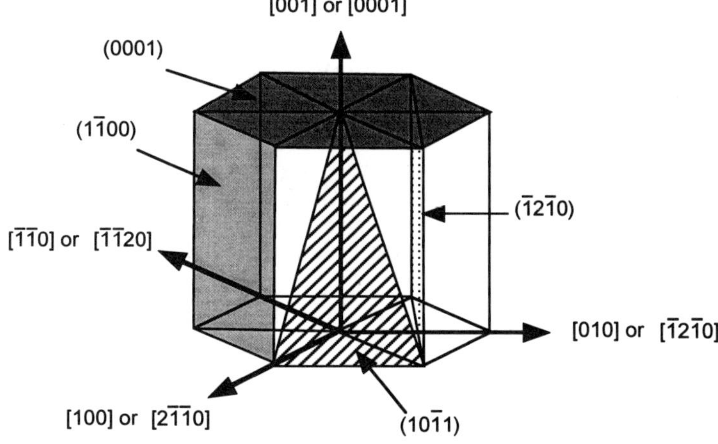

FIG. 35. Indices of planes and directions in the hexagonal system.

2.7. DIFFRACTION FROM CRYSTALLINE MATERIALS—BRAGG'S LAW

In Sec. 1.3 we said that an atom can scatter x-rays, and if many atoms are together then the scattered waves from all the atoms can interfere. If the scattered waves are in phase (coherent), they interfere in a constructive way and we get diffracted beams in specific directions. These directions are governed by the wavelength (λ) of the incident radiation and the nature of the crystalline sample. Bragg's law, formulated by W. L. Bragg in 1913, relates the wavelength of the x-rays to the spacing of the atomic planes.

To derive Bragg's law, we begin by assuming that each plane of atoms partially reflects the incident wave much like a half-silvered mirror. The

Diffraction and Reflection—What's the Difference?

In the derivation of Bragg's law we treat the planes of atoms as mirrors that can reflect the x-rays. This analogy between diffraction and reflection simplifies the derivation: we are all familiar with the reflection of visible light. The angle the incident x-ray beam makes with the lattice planes is the same as that made by the diffracted beam, so there is a qualitative similarity between diffraction of x-rays and reflection of visible light. Diffraction and reflection however, are actually different phenomena. Some important differences are listed here.

- The diffracted beam is composed of waves that have been scattered by all the atoms of the crystal lying in the path of the incident beam. Reflection of visible light is a surface phenomenon occurring in a layer about $\lambda/2$ thick.
- The diffraction of x-rays of a single wavelength takes place only at certain angles, θ (known as the Bragg angle), where there is constructive interference and Bragg's law is thus satisfied. The reflection of visible light takes place at any angle of incidence.
- The intensity of the beam of visible light reflected by a good mirror is almost the same as the intensity of the incident beam. The intensity of a diffracted beam of x-rays from a crystal is extremely small compared to that of the incident beam.
- In reflection, the angle of incidence (or reflection) is defined as the angle between the *normal* to the specimen surface and the incident (or reflected) beam. In diffraction, the angle of the incident beam (or the diffracted one) is defined as the angle between the specimen surface and the incident (or diffracted) beam.

x-rays are not really being reflected—they are being scattered—but it is very convenient to think of them as reflected, and we often call the planes the "reflecting planes" and the diffracted beams the "reflected beams," when we really mean diffracting planes and diffracted beams. Many books use these terms interchangeably, so you'll just have to get used to it. Because of this reflection analogy, the peaks that occur in an x-ray diffraction pattern are often called *reflections*.

Consider the diffracted wave in Fig. 36. It is assumed to make the same angle, θ, with the atomic planes as does the incident wave. Now the criterion for the existence of the diffracted wave is that the scattered ("reflected") x-rays should all be in phase across a wavefront such as BB′. For this to be so, the path lengths between wavefronts AA′ and BB′ for the rays shown must differ by exactly an integral number (n) of wavelengths λ. Therefore the path difference, δ, must be

$$\delta = n\lambda \tag{14}$$

where n is an integer. Now since lines CC′ and CD in Fig. 36 are also wavefronts, we can write

$$\delta = \mathrm{DE} + \mathrm{EC'} = 2\mathrm{EC'} \tag{15}$$

From elementary trigonometry,

$$\delta = 2\mathrm{CE} \sin\theta \tag{16}$$

and because CE is the interplanar spacing d' we can write

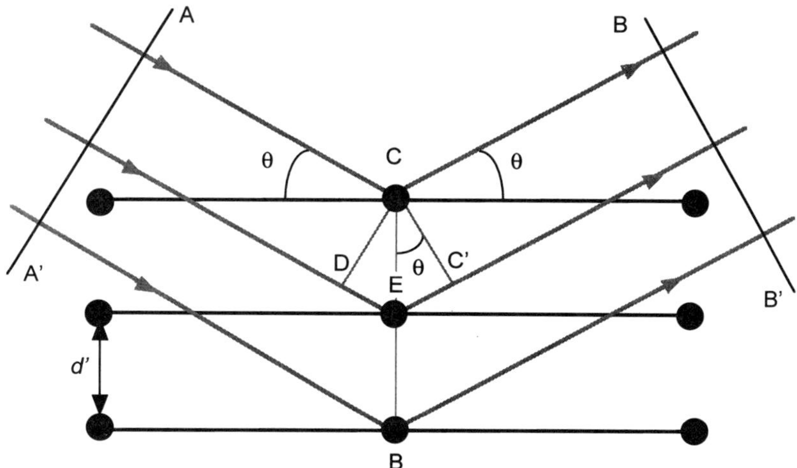

FIG. 36. Diffraction of x-rays by a crystal.

$$\delta = 2d' \sin \theta \tag{17}$$

By combining Eqs. (14) and (17), we get

$$n\lambda = 2d' \sin \theta \tag{18}$$

This equation is known as *Bragg's law* and is extremely important in indexing x-ray diffraction patterns and, hence, for determining things like the crystal structure of the material, as you will see in the experimental modules in Part II. It is well worth committing this equation to memory.

The parameter n is known as the *order of reflection* and is the path difference, in terms of number of wavelengths, between waves scattered by adjacent planes of atoms, as indicated by Eq. (14). A first-order reflection occurs when $n = 1$ and the scattered and incident waves have a path difference of one wavelength. When $n > 1$, the reflections are called higher-order.

We can rewrite Eq. (18) as

$$\lambda = 2\frac{d'}{n} \sin \theta \tag{19}$$

where, as defined in Fig. 36, d' corresponds to the spacing between planes (hkl) and the parameter d'/n corresponds to the spacing between planes $(nh\,nk\,nl)$. So we can consider a higher-order reflection as a first-order reflection from planes spaced at a distance of $1/n$ of the previous spacing. By setting $d = d'/n$ and substituting into Eq. (19), we can write Bragg's law in the customary way as

$$\lambda = 2d \sin \theta \tag{20}$$

As an example, the 440 reflection can be considered a fourth-order reflection from the (110) planes of spacing d' or a second-order reflection from the (220) planes of spacing $d = d'/2$, or even as a first-order reflection from the (440) planes of spacing $d = d'/4$. This convention is consistent with our definition of Miller indices in Sec. 2.6, since the plane with Miller indices $(nh\,nk\,nl)$ is parallel to the plane with Miller indices (hkl) but has a spacing $1/n$ times the spacing of the (hkl) planes. (Note: Higher-order reflections are very important for electron diffraction in the transmission electron microscope.)

2.8. THE STRUCTURE FACTOR

The *structure factor*, F, describes the effect of the crystal structure on the intensity of the diffracted beam. We start by showing the importance of the structure factor in a qualitative way, and then we use a more quantitative approach. Consider the two orthorhombic unit cells in Fig. 37. One cell is base-centered orthorhombic, and the other is body-centered orthorhombic. Both cells have two atoms per cell. Diffraction from the (001) planes of the two cells is shown in Fig. 38. For the base-centered orthorhombic lattice in Fig. 37a, if Bragg's law is satisfied, this means that the path difference *ABC* between wave 1′ and wave 2′ is λ and diffraction will occur. In the same way for the body-centered lattice, if waves 1′ and 2′ are in phase, their path difference is λ. However, another plane of atoms is midway between the (001) planes in the body-centered lattice (Fig. 38b). The path difference between waves 1′ and 3′ is λ/2. These waves will be completely out of phase, destructive interference will occur, and we will get no diffracted beam. A wave reflected from the next plane down will cancel wave 2′, and so on. The net result is that there is no 001 reflection for the body-centered lattice. [Note: The 001 reflection (without brackets) comes from diffraction of the x-ray beam by the (001) planes.] Thus, for certain structures some reflections will be absent from the diffraction pattern. They are known as *forbidden reflections*—they are forbidden by the structure factor. The intensity of the beam diffracted by all the atoms in the unit cell in a given direction predicted by Bragg's law is proportional to F^2. The intensities of the reflections are what we see in the diffraction pattern. If $F^2 = 0$, there is no reflection.

Now we present the equation for calculating F. The derivation of this equation is beyond the scope of this book, but one of the texts in the Bibliography gives the derivation. Anyway,

$$F = \sum_i f_i e^{2\pi i(hu_i + kv_i + lw_i)} \tag{21}$$

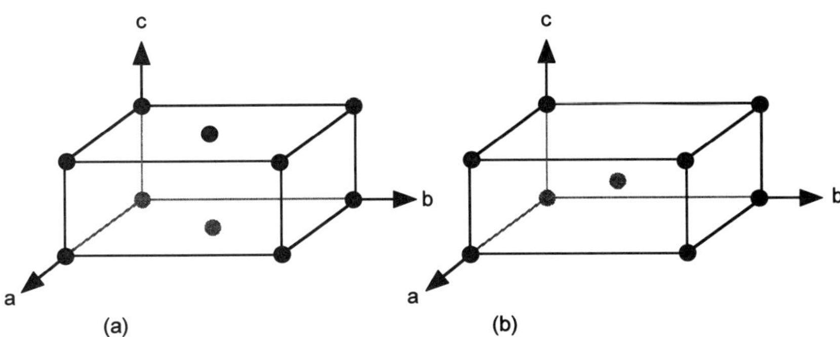

FIG. 37. (a) Base-centered and (b) body-centered orthorhombic unit cells.

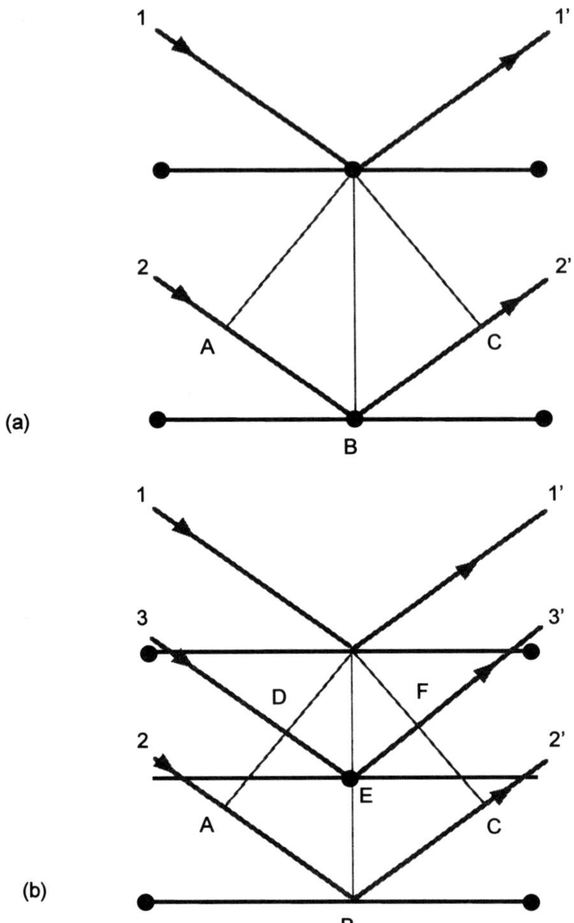

FIG. 38. Diffraction from the (001) planes of (a) base-centered and (b) body-centered orthorhombic unit cells.

where f is the atomic scattering factor, discussed in Sec. 1.3, u, v, and w are the atom positions in the unit cell, and h, k, and l are the Miller indices of the reflection. The summation is performed over all atoms in the unit cell. What does Eq. (21) tell us? Most importantly, it tells us what reflections hkl to expect in a diffraction pattern from a given crystal structure with atoms located at positions u,v,w. These are known as *selection rules*. Reflections for which $F = 0$ will have zero intensity and will not appear in the diffraction pattern; these reflections are called forbidden. Equation (21) is a completely general equation. It applies to *all* crystal lattices regardless of whether there are 1 or 100 atoms per unit cell and whether the structure is cubic, hexagonal, or triclinic.

Useful Relations in Evaluating Complex Exponential Functions

Calculating structure factors using Eq. (21) involves complex exponential functions. The following relations are useful:

$e^{\pi i} = e^{3\pi i} = e^{5\pi i} = -1$
$e^{2\pi i} = e^{4\pi i} = e^{6\pi i} = 1$
$e^{n\pi i} = (-1)^n$, where n is any integer
$e^{n\pi i} = e^{-n\pi i}$, where n is any integer
$e^{ix} + e^{-ix} = 2 \cos x$

Let's do some examples to show how Eq. (21) is used. First consider a simple cubic crystal structure. We know from Sec. 2.4.1 that there is one atom per cell in the simple cubic structure located at 0,0,0. So we can insert the atomic coordinates 0,0,0 for u,v,w, respectively, in Eq. (21). Therefore the structure factor for the simple cubic structure is

$$F = fe^{2\pi i(0)} = f \tag{22}$$

For the simple cubic structure F is independent of h, k, and l, which means that *all* the reflections are allowed. In an x-ray diffraction pattern from a simple cubic material we would see reflections corresponding to beams diffracted by all the lattice planes in the material, such as (100), (110), (111), (200), (210), etc.

Now consider the fcc structure. It has four atoms, of the same type, per unit cell located, as noted in Sec. 2.4.1, at 0,0,0; $\frac{1}{2},\frac{1}{2}$,0; $\frac{1}{2}$,0,$\frac{1}{2}$; and 0,$\frac{1}{2},\frac{1}{2}$. We can therefore insert the atomic coordinates u, v, and w in Eq. (21) and obtain

$$F = fe^{2\pi i(0)} + fe^{2\pi i\left(\frac{h}{2}+\frac{k}{2}\right)} + fe^{2\pi i\left(\frac{h}{2}+\frac{l}{2}\right)} + fe^{2\pi i\left(\frac{k}{2}+\frac{l}{2}\right)} \tag{23}$$

which we can simplify as

$$F = f[1 + e^{\pi i(h+k)} + e^{\pi i(h+l)} + e^{\pi i(k+l)}] \tag{24}$$

If h, k, and l are all even or all odd (we say they are "unmixed"), then the sums $h + k$, $h + l$, and $k + l$ are even integers, and each term in Eq. (24) has the value 1. Therefore for the fcc structure,

$$F = 4f \qquad \text{for unmixed indices}$$

If h, k, and l are mixed (i.e., some are odd and some are even), then the sum of the three exponentials is –1. Let's suppose that $h = 0$, $k = 1$, and $l = 2$; i.e., the plane has indices (012). Then

$$F = f(1 - 1 + 1 - 1) = 0$$

and there is no reflection. Thus,

$$F = 0 \quad \text{for mixed indices}$$

Hence, for a material with the fcc structure we will see reflections corresponding to planes such as (111), (200), and (220), but we will not see reflections for planes (100), (110), (210), (211), etc.

In our two examples we chose cubic unit cells. *The structure factor is independent of the shape and size of the unit cell.* Any primitive cell will show reflections corresponding to diffraction from all the lattice planes; any face-centered cell will show reflections only when h, k, and l are unmixed. However, not all unmixed reflections may be seen in all face-centered cells; some may be absent depending on the number of atoms present in the unit cell. What we can be certain of is that no reflection will be observed when h, k, and l are mixed.

Now let's consider unit cells containing a basis of two atoms per lattice point. The NaCl structure, illustrated in Fig. 26, is built on the face-centered cubic (cubic F) Bravais lattice with a basis of two atoms, or, to be more accurate, ions (one Na and one Cl), per lattice point. To calculate the structure factor for NaCl, we can use the result for the fcc structure and include the contribution of the basis. The NaCl structure has eight atoms per unit cell, four Na and four Cl. (That we actually have ions instead of atoms makes little difference when determining F since we generally don't take into account the charge of the ion.) The atomic scattering factors for atoms and ions are slightly different at small angles, but at higher angles they are almost identical. In Table 7 we list the values of f as a function of $(\sin\theta)/\lambda$ for Na, Na^+, Cl, and Cl^-.

Normally, for NaCl, we take the basis to consist of one Na located at 0,0,0 and one Cl located at $\frac{1}{2}$,0,0 (as we did in Sec. 2.4.3). We could just as well take the basis as the Na ion at 0,0,0 and the Cl ion at $\frac{1}{2},\frac{1}{2},\frac{1}{2}$. Look at Fig. 26 to see why this is so. We can write, using Eq. (21), the structure factor for the NaCl structure as

TABLE 7. Atomic Scattering Factors for Na, Na^+, Cl, and Cl^-

$(\sin\theta)/\lambda$ (nm^{-1})	0.0	1.0	2.0	3.0	4.0	5.0	6.0	7.0	8.0	9.0	10.0
Na^+	10	9.546	8.374	6.894	5.471	4.290	3.395	2.753	2.305	1.997	1.785
Na	11	9.760	8.335	6.881	5.471	4.293	3.398	2.754	2.305	1.997	1.784
Cl	17	15.23	11.99	9.576	8.181	7.305	6.595	5.915	5.245	4.607	4.023
Cl^-	18	15.69	11.99	9.524	8.162	7.305	6.600	5.920	5.248	4.608	4.024

$$F = \left[f_{\text{Na}} e^{2\pi i(0)} + f_{\text{Cl}} e^{2\pi i\left(\frac{h}{2}+\frac{k}{2}+\frac{l}{2}\right)} \right] \left[1 + e^{\pi i(h+k)} + e^{\pi i(h+l)} + e^{\pi i(k+l)} \right] \tag{25}$$

Simplifying gives

$$F = \left[f_{\text{Na}} + f_{\text{Cl}} e^{\pi i(h+k+l)} \right] \left[1 + e^{\pi i(h+k)} + e^{\pi i(h+l)} + e^{\pi i(k+l)} \right] \tag{26}$$

The first term in Eq. (26) represents the basis of the unit cell, the Na ion at 0,0,0 and the Cl ion at $\frac{1}{2},\frac{1}{2},\frac{1}{2}$. The second term is the same as the term in brackets in Eq. (24); it is the structure factor for the fcc structure. Unlike the calculation of the structure factor for a material with the fcc structure, for NaCl we have to include the atomic scattering factors for the different types of atoms. The selection rules for the NaCl structure are

$$\begin{aligned} F &= 4(f_{\text{Na}} + f_{\text{Cl}}) \quad \text{if } h, k, l \text{ are even} \\ F &= 4(f_{\text{Na}} - f_{\text{Cl}}) \quad \text{if } h, k, l \text{ are odd} \\ F &= 0 \quad \text{if } h, k, l \text{ are mixed} \end{aligned}$$

The third condition is the same for the fcc structure (this shows that NaCl has the face-centered cubic Bravais lattice). The *first two* conditions are new and depend on $f_{\text{Na}} \pm f_{\text{Cl}}$. That the NaCl structure contains more than four atoms per unit cell has not led to elimination of any reflections seen for the fcc structure, but it has led to a change in the intensity of the reflections. For example, the 111 reflection for NaCl will be weaker than the 200 reflection because the former involves the difference, rather than the sum, of the atomic scattering factors of the atoms. (In the next example we will see a case of elimination of reflections.) These intensity calculations help us determine atom positions in a crystal.

You can also arrive at the same set of selection rules for the NaCl structure by doing the determination long-hand. To do this, substitute the atom positions for the eight atoms in the unit cell of the NaCl structure into Eq. (21) and simplify the equation. But whenever a lattice contains a basis, the equation for the structure factor can be written out quickly. The first term is the atom positions in the basis, and the second term is that for the Bravais lattice of the crystal structure.

We will do one more example to illustrate the point we just made; we will determine F for the diamond cubic structure. The diamond cubic structure is based on the face-centered cubic Bravais lattice with a basis of two atoms, giving a total of eight atoms per unit cell. The atoms in the basis are located at 0,0,0 and $\frac{1}{4},\frac{1}{4},\frac{1}{4}$. Both atoms are of the same type (in diamond they are all carbon atoms), so

$$F = \left[f_{\text{C}} + f_{\text{C}} e^{\left(\frac{\pi}{2}\right)i(h+k+l)} \right] \left[1 + e^{\pi i(h+k)} + e^{\pi i(h+l)} + e^{\pi i(k+l)} \right] \tag{27}$$

The first term represents the basis, and the second term the face-centered cubic Bravais lattice. The selection rules for the diamond cubic structure are a little more complicated:

$F = 0$	if h, k, l are mixed (same condition as for the fcc structure)
$F = 4f_c(1 + e^{(\frac{\pi}{2})i(h+k+l)})$	if h, k, l are odd
$F = 0$	if h, k, l are even and $h + k + l = 2N$, where N is odd (e.g., the 200 reflection)
$F = 8f_C$	if h, k, l are even and $h + k + l = 2N$, where N is even (e.g., the 400 reflection)

Once again, $F = 0$ if h, k, and l are mixed. This is the same condition for the fcc structure, but it applies to the diamond cubic structure because it has the same Bravais lattice as the fcc. If all indices are odd, the expression is complex. However, remember that in the diffraction pattern we see the intensities of the reflections (i.e., F^2 not F), so we consider F^2, which is $32f_C^2$.

The Lorentz–Polarization Factor

The Lorentz–polarization factor is actually a combination of two factors—the Lorentz factor and the polarization factor—that influence the intensity of the diffracted beam. The origin of the polarization factor is that the incident beam from the x-ray source is unpolarized. It may be resolved into two plane-polarized components, and the total scattered intensity is the sum of the intensities of these two components, which depends on the scattering angle (i.e., 2θ).

The Lorentz factor takes into account certain geometrical factors related to the orientation of the reflecting planes in the crystal, which also affect the intensity of the diffracted beam.

$$\text{Lorentz–polarization factor} = \frac{1 + \cos^2 2\theta}{\sin^2 \theta \cos \theta}$$

Values of this factor are given in Appendix 7. The Lorentz–polarization factor varies strongly with Bragg angle, θ, and the overall effect is that the intensity of reflections at intermediate Bragg angles is decreased compared to those at high or low angles. The Lorentz–polarization factor is important in all intensity calculations, and we will use it in Experimental Module 7.

Let's now briefly summarize some of the similarities and differences between the NaCl structure and the diamond cubic structure. Both structures are based on the face-centered cubic Bravais lattice and have eight atoms per unit cell. However, because the atom positions are different in the two structures the x-ray diffraction patterns are different: there are different reflections and different intensities. The absence of some reflections and the differences in intensities (F^2) give us a way to find the atom positions in the lattice. However, in the experimental modules in Part II we do not emphasize this aspect.

As an exercise, what do you think the selection rules would be for the zinc blende structure? The zinc blende structure is similar to the diamond cubic, but the basis consists of two different types of atom, one S at 0,0,0 and one Zn at $\frac{1}{4},\frac{1}{4},\frac{1}{4}$. (The answer is given in Table 8.)

Table 8 summarizes the results of structure factor calculations. Note that the intensity of a diffracted beam depends not only on F^2 but on other factors also. The integrated intensity, I, of a diffracted beam is

$$I = F^2 p \left(\frac{1 + \cos^2 2\theta}{\sin^2 \theta \cos \theta} \right) e^{-2M} \tag{28}$$

where p is the multiplicity factor, the term in parentheses is the Lorentz–polarization factor, and e^{-2M} is the temperature factor.

TABLE 8. Examples of Selection Rules for Different Crystal Structures

Crystal Type	Bravais Lattice	Reflections Present for	Reflections Absent for
Simple	Primitive	Any h, k, l	None
Body-centered	Body-centered	$h + k + l$ even	$h + k + l$ odd
Face-centered	Face-centered	$h, k,$ and l unmixed	$h, k,$ and l mixed
NaCl	Face-centered cubic	$h, k,$ and l unmixed	$h, k,$ and l mixed
Zinc blende	Face-centered cubic	$h, k,$ and l unmixed	$h, k,$ and l mixed
Diamond cubic	Face-centered cubic	As fcc, but if all even and $h + k + l \neq 4N$; then absent	$h, k,$ and l mixed and if all even and $h + k + l \neq 4N$
Base-centered	Base-centered	h and k both even or both odd[a]	h and k mixed[a]
Hexagonal close-packed	Hexagonal	$h + 2k = 3N$ with l even $h + 2k = 3N \pm 1$ with l odd $h + 2k = 3N \pm 1$ ωιτη l even	$h + 2k = 3N$ with l odd

[a]These relations apply to a cell centered on the C face. If reflections are present only when h and k are unmixed, or when k and l are unmixed, then the cell is centered on the B or A face, respectively.

The Temperature Factor

The temperature factor takes into consideration the effect of temperature on the intensity of the reflections in the x-ray diffraction pattern. At all temperatures, even at the absolute zero of temperature, the atoms in a crystal are undergoing thermal vibration about their mean positions. One of the effects of this vibration is that the intensities of the reflections decrease as the temperature increases. It is easy to see why. We normally view each lattice plane as consisting of atoms in very well defined positions and the planes themselves are perfectly flat and of uniform thickness. Thermal vibration of the atoms causes the spacing of the planes to become rather ill defined. Diffraction by sets of parallel planes at the Bragg angle is no longer perfect, as it would be if the atoms were not moving.

If the average displacement of an atom from its mean position is u, the intensity of the reflections decreases as u increases (i.e., as the temperature increases). The value of u depends on the elastic constants of the crystal—the stiffer the crystal the lower the value of u.

At a fixed temperature (say room temperature) the intensity decreases as θ increases because at large Bragg angles planes with low values of d are diffracting. Hence, high-angle reflections are decreased in intensity relative to low-angle reflections. This effect is taken into account by the temperature factor e^{-2M}, where M is a function of several variables, including u and the diffraction angle 2θ.

Calculation of the temperature factor is quite complicated, and the data required to do this calculation are not readily available for nonmetallic materials and alloys. For this reason we will not use the temperature factor in our calculation of intensities in Experimental Module 7.

2.9. DIFFRACTION FROM AMORPHOUS MATERIALS

X-ray diffraction can also be used to provide information about the structure of amorphous or noncrystalline materials such as glass. In fact, it was early x-ray diffraction work by Warren and his colleagues that led to some of the early theories of glass structure. X-ray diffraction patterns from amorphous SiO_2 (silica glass) and a crystalline form of SiO_2 (cristobalite) are shown in Fig. 39. The crystalline material shows a series of sharp peaks, or reflections, due to diffracted beams arising from different lattice planes. The glass shows one broad peak centered in the range in which the strong peak was seen in the diffraction pattern of the crystalline material of the same composition. This observation led to the conclusion that in an amorphous material there is no long-range order as there is in a crystal. The amorphous material only exhibits short-range order. You

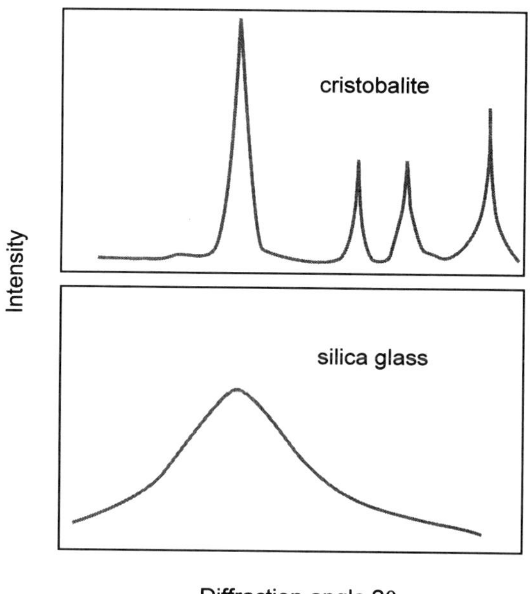

FIG. 39. X-ray diffraction patterns from (a) crystalline cristobalite and (b) silica glass.

will find out in Experimental Module 6 that broadening of the x-ray peaks occurs when the crystallite or grain size of the material is very small (<0.1 μm). Peak broadening increases with decreasing crystallite size. Figure 40 illustrates the difference between (a) an ordered crystalline form and (b) a possible structure of a glassy form of the same composition.

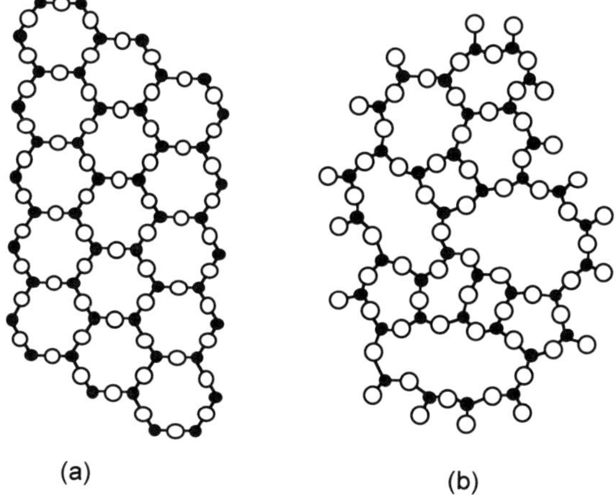

FIG. 40. Schematic representation of (a) a crystalline form and (b) an amorphous form of the same composition.

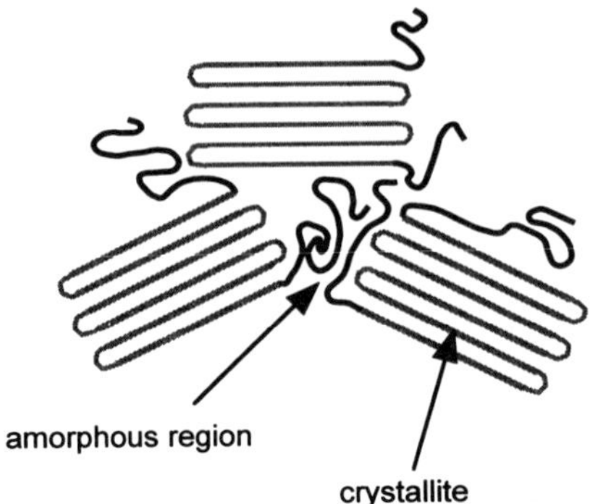

FIG. 41. Crystallites in a polymer.

Polymers are composed of long molecular chains normally arranged in a random way. In some polymers, particularly those with simple linear conformations, the chains can become organized into ordered regions as shown in Fig. 41. These ordered regions are called crystallites. Many polymers are partly crystalline; they have crystallites dispersed in an otherwise amorphous matrix. For example, polyethylene (PE) consists of molecular chains of composition $-(CH_2-CH_2)_n-$ and polytetrafluoroethylene (PTFE) has molecular chains of composition $-(CF_2-CF_2)_n-$ and both show high volume fractions of crystallites. The size of the crystallites in a partly crystalline polymer can be determined by x-ray diffraction by measuring the broadening of the diffraction peaks.

3

Practical Aspects of X-Ray Diffraction

3.1. GEOMETRY OF AN X-RAY DIFFRACTOMETER

In this and the following sections we provide some general experimental background to the x-ray diffraction technique. Each manufacturer's instrument has its unique features, and for specific details you should refer to the instrument operating manual that came with your diffractometer. The experimental geometry used in the powder diffraction method is illustrated in Fig. 42. The three basic components of an x-ray diffractometer are the

- X ray source
- Specimen
- X ray detector

and they all lie on the circumference of a circle, which is known as the *focusing circle*. The angle between the plane of the specimen and the x-ray source is θ, the Bragg angle. The angle between the projection of the x-ray source and the detector is 2θ. For this reason the x-ray diffraction patterns produced with this geometry are often known as θ–2θ (theta–two theta) scans. In the θ–2θ geometry the x-ray source is fixed, and the detector moves through a range of angles. The radius of the focusing circle is not constant but increases as the angle 2θ decreases, as you can see from Fig. 42. The 2θ measurement range is typically from 0° to about 170°. In an experiment you need not necessarily scan the whole range of detector angles. A 2θ range from 30° to 140° is an example of a typical scan. The choice of range depends on the crystal structure of the material

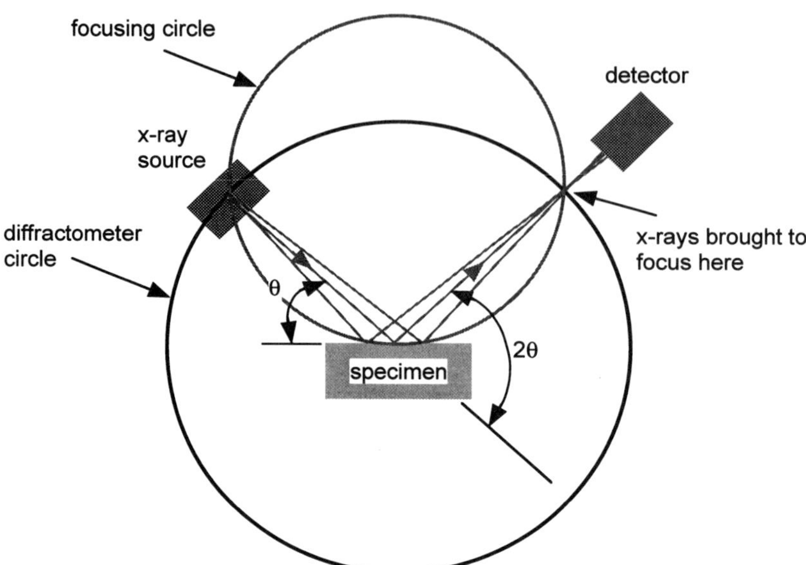

FIG. 42. Geometry of an x-ray diffractometer.

(if known) and the time you want to spend obtaining the diffraction pattern. For an unknown specimen a large range of angles is often used because the positions of the reflections are not known, at least not yet!

The θ–2θ geometry is the most common, but you may come across other geometries. In the θ–θ (theta–theta) geometry both the x-ray source and the detector move in the vertical plane in opposite directions above the center of the specimen. In some forms of x-ray diffraction analysis the sample is tilted about an axis ψ (psi). This scan, often known as an ω (omega) scan, provides a measure of the strain (change in length divided by original length, $\Delta l/l$) in the specimen, from which it is then possible to determine the stress (force per unit area, F/A). The specimen can be rotated about the ψ axis and about an orthogonal axis called the ϕ (phi) axis. This type of scan, sometimes called a ϕ scan, is particularly useful for characterizing oriented polycrystalline thin films where the rotation axis of the grain boundaries is often predominantly aligned perpendicular to the surface of the film. The ω scan and ϕ scan are very specialized techniques, yet, although very useful, are not widely used. In the remainder of this chapter we consider only the θ–2θ geometry.

The *diffractometer circle* in Fig. 42 is different from the focusing circle. The diffractometer circle is centered at the specimen, and both the x-ray source and the detector lie on the circumference of the circle. The radius of the diffractometer circle is fixed. The diffractometer circle is also referred to as the *goniometer circle*. The goniometer is the central component of an x-ray diffractometer and contains the specimen holder. It has arms to which the x-ray source and the detector are mounted. In most powder diffractometers the goniometer is vertically mounted, but in other diffractometers, for example those used to study thin films, the goniometer may be horizontally mounted.

3.2. COMPONENTS OF AN X-RAY DIFFRACTOMETER

3.2.1. The X-Ray Source

X-rays are generated, as described in Sec. 1.2, by directing an electron beam of high voltage at a metal target anode inside an evacuated x-ray tube. The voltage and current are both variables, and you should check the operating manual or with the instrument operator or technician to see what values you should use. For all of the experimental modules in Part II use whatever values of voltage and current are recommended for your instrument. Suitable operating voltages depend on the target metal (remember that they should be above the critical excitation potential for knocking out K electrons), e.g., molybdenum 50 to 55 kV, copper 25 to 40 kV, iron 25 to 30 kV, and chromium 25 kV. Copper is the most frequently used target, and typical operating conditions are 40 kV and 30 mA.

The choice of radiation depends on several factors, including how penetrating it is (we do not want the x-rays absorbed by the specimen). Recall from Sec. 1.2 that when an incident electron has sufficient energy it can knock out an electron from the K shell of an atom. When the atom returns to the ground state, an x-ray is emitted (a characteristic x-ray). In a similar way, an incident x-ray, with enough energy, can also knock out an electron from the K shell of the atom. The atom is now in an excited state, and an electron in a higher energy level then fills the hole in the K shell, to return the atom to its ground state, with the emission of an x-ray. The emitted radiation in this case is known as *fluorescent radiation*. It radiates in all directions and has exactly the same wavelength as the characteristic radiation.

If an element in the specimen has an atomic number slightly below that of the target metal, the x-rays will be strongly absorbed. This results in a decrease in the intensity of the diffracted beam and a large amount of fluorescent radiation, which increases the background signal. For example, if we are using Cu $K\alpha$ x-rays, which have an energy of 8.04 keV,

Absorption

Radiation incident upon a material is absorbed when electrons in the atoms that make up the material are excited from a lower energy state (usually the ground state) to a higher energy state. In a solid, consisting of many atoms, the energy levels are in the form of bands. When we are talking about electronic applications, we refer to the outer two bands as the valence band and the conduction band. But when we are talking about optical properties of materials it is common to refer to the two bands as the valence band and the absorption band, respectively. The selective absorption of radiation gives rise to color in many materials.

to examine a specimen containing iron (atomic number 26), the x-rays have enough energy to eject a K shell electron from iron (the critical ionization energy for a K shell electron in iron is about 7 keV). The result is the emission of fluorescent radiation. To avoid fluorescence from specimens containing iron and other transition elements, Cr Kα radiation, which has an energy of 5.41 keV (lower than the critical ionization energy for a K shell electron in iron), may be used.

As noted in Sec. 1.2, the intensity of the characteristic lines in the x-ray spectrum increases with an increase in the applied voltage. But the values of current and voltage will not affect the positions (2θ values) of the reflections in the diffraction pattern. The load (product of voltage and current) is usually kept just under the maximum rating (typically around 2 kW) for the x-ray tube in order to make it last longer.

In most diffractometers a fixed value of the x-ray wavelength λ, is used. For example, in the experiments described in the experimental modules we have assumed that Cu Kα radiation will be used, which has a weighted average wavelength $\lambda = 0.154184$ nm. Some of the commonly used K wavelengths were listed in Table 2, but Cu Kα is generally the most useful because of its high intensity.

3.2.2. The Specimen

In a typical x-ray diffraction experiment, a thin layer of crystalline powder is spread onto a planar substrate, which is often a nondiffracting material such as a glass microscope slide, and exposed to the x-ray beam. You will use this procedure to complete the experimental modules in Part II. The quantity of powder used for each experiment is quite small, usually a few milligrams (1 mg = 10^{-3} g).

Ideally, the specimen should contain numerous small, equiaxed randomly oriented grains. The grain size of the powders should be ≤50 μm; i.e., the powder can pass through a U.S. standard 325 mesh sieve. For materials with high symmetry, such as cubic or hexagonal, the grain size of the powder can be up to about 70 μm, because of the multiplicity factor. If you refer to Sec. 2.6 you will see why the multiplicity factor becomes important for crystals with high symmetry. In such crystals several planes can contribute to each reflection; e.g., diffraction from the (100), (010), (001), ($\bar{1}00$), ($0\bar{1}0$), and ($00\bar{1}$) planes in a cubic crystal all contribute to the same reflection (the 100 reflection).

When the grains are very small, <1 μm, broadening of the peaks in the diffraction pattern occurs, as illustrated in Experimental Module 6. If the grains in the specimen are too large, then effects associated with preferred orientation may become significant. Later in this section we describe what these effects are.

It is important that the specimen you choose for analysis be representative of the entire sample. For example, if you are going to characterize a mixture of two powders (as in Experimental Module 7) or if a standard has been added (as in Experimental Modules 6 and 7), be sure that the two powders are thoroughly mixed, otherwise the small amount you take for x-ray diffraction analysis may not contain the two powders in the same proportion as in the global mixture. In addition to using powders, the specimen may also be a sheet or film of a polycrystalline material, for example a sheet of aluminum or a polycrystalline film on a substrate (e.g., a TiN film on steel).

As noted, in a θ–2θ scan the position of the incident x-ray beam with respect to the specimen is fixed. In a powder or a polycrystalline material in general, the grains are often randomly oriented, and some grains will always be oriented in a favorable direction with respect to the x-ray beam, to allow diffraction to occur from a specific set of lattice planes. Each set of lattice planes in the crystal having spacings d_1, d_2, d_3, . . . , will diffract at different angles θ_1, θ_2, θ_3, . . . , where θ increases as d decreases in such a way as to satisfy Bragg's law. The intensity of the diffracted beam at each of these different angles is detected, and this is what forms the x-ray diffraction pattern.

If the specimen is textured (i.e., there is a preferred grain orientation) then some of the reflections in the x-ray diffraction pattern may be anomalously intense or some may be absent. In other words, there may be a change in the relative intensities of the reflections compared to those obtained with a randomly oriented powder specimen. By comparing the relative intensities of the reflections from the textured specimen with the random powder pattern, we can determine the degree of preferred orientation. Texturing is common in materials that have a layered structure with easy cleavage between the layers (e.g., mica), thin films, and heavily cold-worked metals. The reflections missing in a diffraction pattern obtained from a textured material are missing not because they are forbidden by the structure factor but because the grains are not oriented in the correct way to allow diffraction to occur from these planes. For example, in a cubic material in which the [100] direction of all the grains is oriented perpendicular to the surface of the specimen, diffraction occurs from the (100) planes but not from the (001) or the (010) planes. Therefore, we would expect the relative intensity of the 100 reflection to be weaker in the textured specimen than in the randomly oriented specimen.

Most specimens are flat, and as you can see in Fig. 42 the face of a flat specimen forms a tangent to the focusing circle. However, for perfect focusing the specimen should be curved to fit the circumference of the focusing circle. For most specimens it is not possible to meet this requirement. The flat specimen causes some broadening of the diffracted beam and can result in a small shift in peak position toward smaller angles, particularly at 2θ angles less than about 60°. These effects can be reduced by decreasing the horizontal divergence (the spread or width) of the incident beam, but this, in turn, leads to a decrease in intensity.

3.2.3. The Optics

On the x-ray source side, a line source of x-rays passes through a series of slits called Soller slits. These slits consist of a series of closely spaced parallel metal plates that define and collimate (i.e., make it parallel) the incident beam. Soller slits are typically about 30 mm long and 0.05 mm thick, and the distance between the plates is about 0.5 mm. The slits are usually made of a metal with a high atomic number, such as molybdenum or tantalum (because of their high absorption capacity). The arrangement of slits in an x-ray diffractometer is illustrated in Fig. 43. The divergence slit on the incident beam side defines the divergence (or width) of the incident beam.

After the beam has been diffracted by the specimen, it passes through another set of slits, as shown in Fig. 43. The antiscatter slit reduces the background radiation, improving the peak-to-background ratio, by making sure that the detector can receive x-rays only from the specimen area. The beam converges on passing the receiving slit, which defines the width of the beam admitted to the detector. An increase in slit width increases the maximum intensity of the reflections in the diffraction pattern but

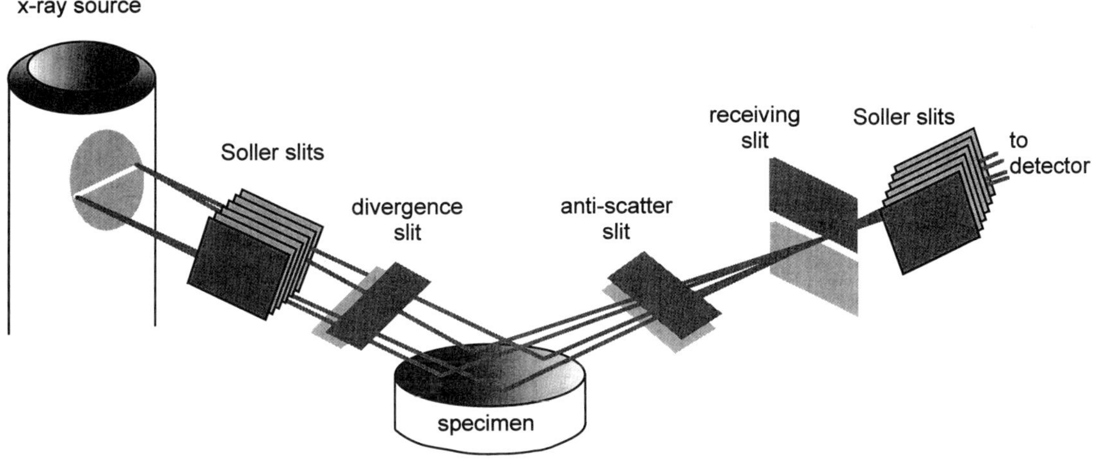

FIG. 43. Arrangement of slits in a diffractometer.

generally results in some loss of resolution. However, the integrated intensity, *I*, of a reflection is independent of slit width. The integrated intensity is the peak area as shown in Fig. 44. A change in slit width does not change the ratio of integrated intensities of two peaks, I_1/I_2, but it most

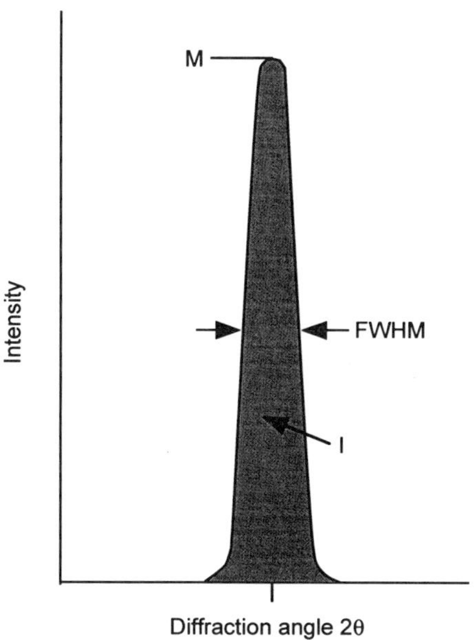

FIG. 44. Typical peak in an x-ray diffraction pattern.

likely will change the ratio of the maximum intensities M_1/M_2. That is why we always use the integrated intensity. Furthermore, because it is difficult to measure absolute intensities, we always take the relative integrated intensities of the reflections. This is what we use in all the experimental modules in Part II.

Another set of Soller slits is placed after the receiving slit on the diffracted beam side, before the monochromator. We now describe the function of the monochromator.

To achieve monochromatic radiation, we must remove unwanted radiation, say the Kβ radiation. In early x-ray diffractometers this was achieved with a filter that strongly absorbed the Kβ radiation. In most modern diffractometers a monochromating graphite crystal is used. The graphite monochromator is usually placed in the diffracted beam, i.e., after the specimen (Fig. 45) because it suppresses the background radiation (such as fluorescent radiation) originating from the specimen. The monochromator is oriented in such a way that it diffracts *only* the Kα radiation. The beam still consists of the wavelengths Kα_1 and Kα_2. Although it is possible to further isolate the Kα_1 component from the Kα_2 component by using a high-quality monochromator placed in the path of the incident beam (i.e., before the specimen), in most cases this is not done because there may be a loss in beam intensity. Further, it is often very useful to have both components in the x-ray diffraction pattern, as you will see in Experimental Module 3 when we demonstrate how to obtain precise measurements of the lattice parameter. If a monochromator

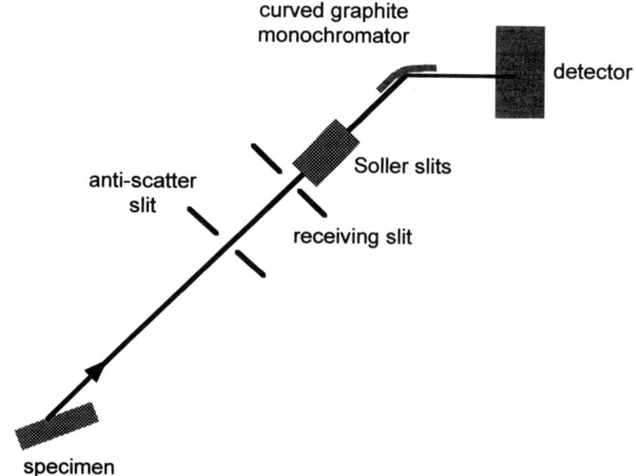

FIG. 45. Schematic of a diffractometer with a monochromating crystal in the diffracted beam.

is used, the set of Soller slits on the diffracted beam side, situated just before the monochromator, is not essential because the monochromating crystal reduces the divergence of the beam.

There has been, and continues to be, considerable research and development activities in x-ray optics. These activities are directed primarily toward collimating or focusing x-rays. For x-ray diffraction we require a well-collimated beam of x-rays. The approach discussed so far to achieve this uses a slit to eliminate x-rays not traveling in the preferred direction. The limitation of this approach is that it rejects most of the output from the x-ray source and results in a relatively low flux of x-rays delivered to the specimen. One method, first proposed in 1986, to capture a large amount of x-rays from the source and deliver a significant proportion of them to the specimen is the use of a special type of optical fiber called a polycapillary optical fiber. Each optical fiber consists of hundreds or even thousands of tiny hollow glass capillaries. The outer diameter of the fibers varies from 300 to 600 μm, and each capillary has an inner diameter between 3 and 50 μm. The hollow glass capillaries act as waveguides, and the x-rays undergo total reflection at the capillary surface, as shown in Fig. 46. In a capillary of length 100 mm, the photon may undergo as many as 100 reflections. After several years of research and development, these polycapillary optical fibers are now commercially available.

Another method developed for collimating and focusing x-rays uses a mirror consisting of a series of layers (called a multilayer) (Fig. 47). Typically the mirror has 50 layers each of tungsten and silicon. For

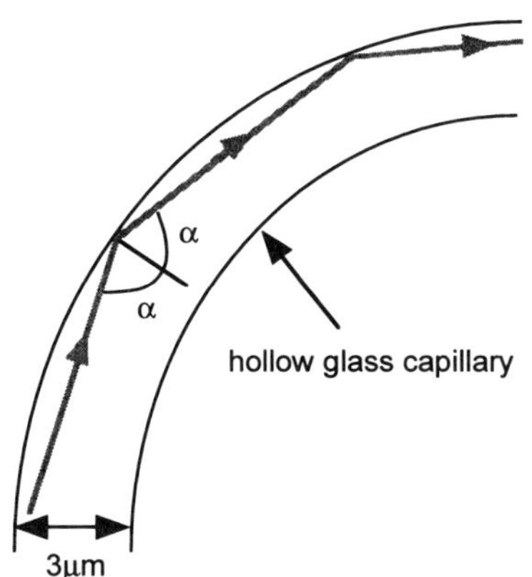

FIG. 46. X-ray beam undergoing multiple total reflections in a hollow glass capillary.

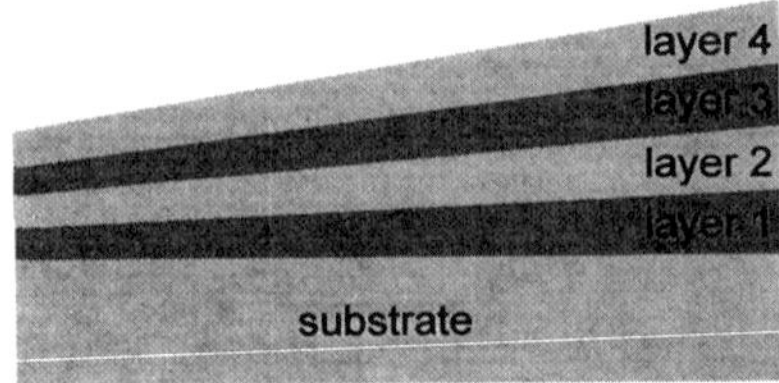

FIG. 47. Schematic of multilayer mirror.

collimating x-rays the mirror is parabolic, and for focusing x-rays the mirror is elliptical. The thickness of the layers is not uniform across the substrate, and this is the key to why these mirrors work. The incident angle of a divergent beam of x-rays onto a parabolic or elliptical surface changes from point to point along the length of the mirror. If the multilayer coating was uniform across the substrate surface, it would not reflect the same wavelength of radiation across its surface. Careful grading of the spacing of the layers makes it possible to collect a large solid angle from an x-ray source. A further advantage of these multilayer mirrors is that they render the beam monochromatic. Parallel beam x-ray diffraction using a parabolic mirror on the incident beam side has particular application in the analysis of irregularly shaped specimens, transparent materials, and high-resolution x-ray diffraction.

3.2.4. The Detector

Three main types of x-ray detector are used on x-ray diffractometers:

1. Proportional
2. Scintillation
3. Solid-state

To determine which type of detector is used on your apparatus, consult the equipment manual or ask the technician. In many new instruments, particularly those dedicated to powder work, a proportional detector is probably used. Scintillation detectors, although common on early instruments and still available, are not widely installed on new diffractometers. The solid-state detector offers many advantages, as we will see, but its high cost is often a factor against its use. Although you don't need to know how the detector works in order to use it, a basic understanding will help you realize why certain experimental procedures are adopted.

Proportional Detector

The gas proportional detector (Fig. 48) is the most commonly used. It consists of a tube filled with a noble gas (e.g., xenon) and has a thin positively charged tungsten wire running down the center. When an x-ray photon enters the tube through a thin window, it is absorbed by an atom of the gas and causes a photoelectron to be ejected. A photoelectron is simply an electron produced by ionization of an atom by a photon. The ionization energy of the noble gases is about 30 eV; one copper x-ray photon (8.04 keV) gives rise to about 270 electron–ion pairs. The photoelectron loses its energy by ionizing other gas atoms. The released electrons are attracted to the positively charged tungsten wire, giving rise to a charge pulse. For each x-ray photon entering the detector a few hundred electrons are created. This produces only a small amount of charge, which needs further amplification. However, if the positive potential on the tungsten wire is large enough (about 1 kV), then the electrons produced as a result of the initial x-ray photon are accelerated. When they collide with other gas atoms, they produce further ionization (known as secondary ionization or "avalanche" production), which results in a large increase in the number of electrons and, hence, an increase in the charge by several orders of magnitude. The collected charge (the size of the pulse) is proportional to the energy of the incident photon, and hence the term "proportional detector." The proportionality is important because it allows us to distinguish between x-ray photons with different energies (wavelengths).

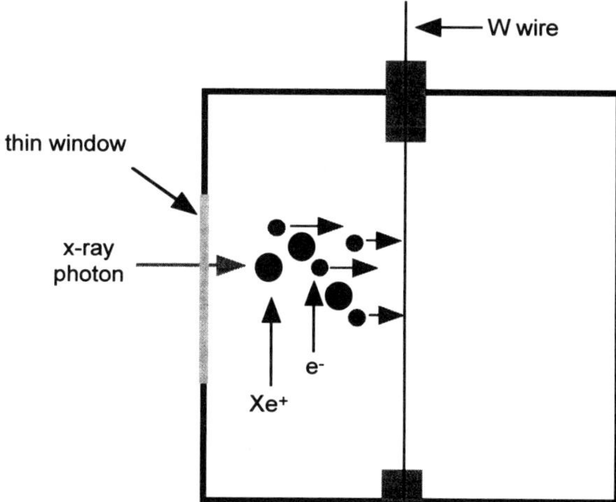

FIG. 48. Gas proportional detector.

Fluorescence

Fluorescence is the emission of electromagnetic radiation, normally visible light, from a material when it goes from an excited state, in which it has absorbed energy, to a lower energy state. In fluorescence, this transition occurs very rapidly (within 10 ns). Figure 50 shows an energy-band diagram for a material with a band gap (an energy gap E_g) between the valence and conduction bands. Absorption of a photon with energy equal to the energy gap causes an electron in the valence band to be excited into the conduction band. When the electron falls back into the valence band, a photon with energy E_g is emitted. If E_g is in the visible part of the electromagnetic spectrum, visible light is emitted. In the NaI:Tl^+ crystal described in this section, the x-rays excite some of the electrons from the valence band to the conduction band in NaI. The alkali halides have large band gaps and so do not produce fluorescence; however, these excited electrons transfer some of their energy to the electrons in the Tl^+ ion. When the Tl^+ ion returns to its ground state, light is emitted. The difference between the energy levels in the Tl^+ ion corresponds to light in the visible part of the electromagnetic spectrum.

In a position-sensitive proportional detector, the tungsten wire is long, and the time required for the pulse to travel from the point of impact of the electrons with the wire to the end of the wire is used to determine the position of the x-ray.

Scintillation Detector

In a scintillation detector (Fig. 49) the incident x-rays cause a crystal (for example, sodium iodide doped with 1% thallium) to fluoresce. For an NaI crystal doped with Tl^+ the fluorescence is in the violet part ($\lambda \approx 420$ nm) of the electromagnetic spectrum. A flash of light (a scintillation) is produced in the crystal for every x-ray photon absorbed. The amount of light emitted is proportional to the x-ray intensity and can be measured by a photomultiplier. The size of the pulses is proportional to the energy of the x-ray photon absorbed.

Although, the scintillation detector is very efficient over a range of x-ray wavelengths, its energy resolution is not as good as the proportional or solid-state detectors (Fig. 51). The energy resolution is determined by the peak widths: the narrower the peaks the better the energy resolution. As a result, it is more difficult to discriminate, using a scintillation detector, between x-ray photons of different wavelengths (energies) on the basis

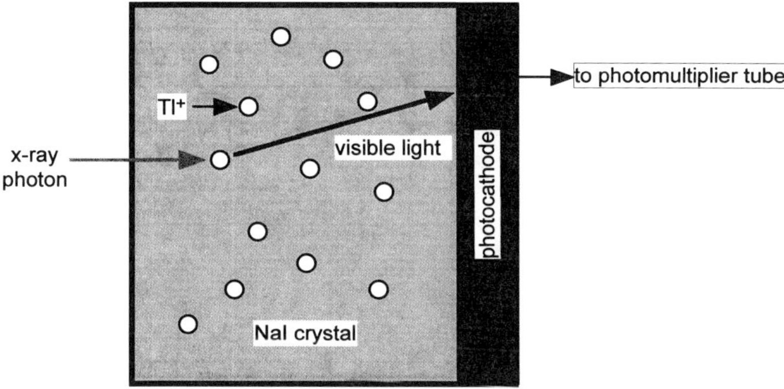

FIG. 49. Scintillation detector.

of pulse size. Scintillation detectors are rare on modern x-ray diffractometers.

Solid-State Detector

Figure 52 is a schematic of a solid-state detector. The detector is a single crystal consisting of a sandwich of intrinsic (pure) silicon between a *p*-type layer (where holes are the majority charge carriers) and an *n*-type layer (where electrons are the majority charge carriers), forming a *p–i–n* (*p*-type–*i*ntrinsic–*n*-type) diode. Intrinsic silicon is essential for the detec-

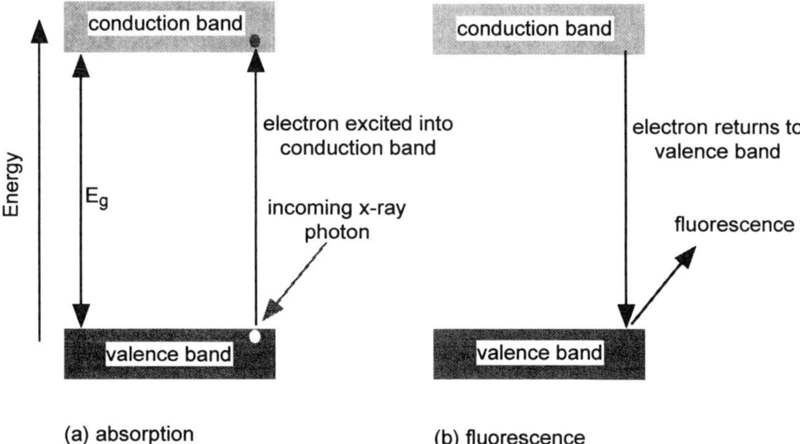

FIG. 50. Illustration of fluorescence.

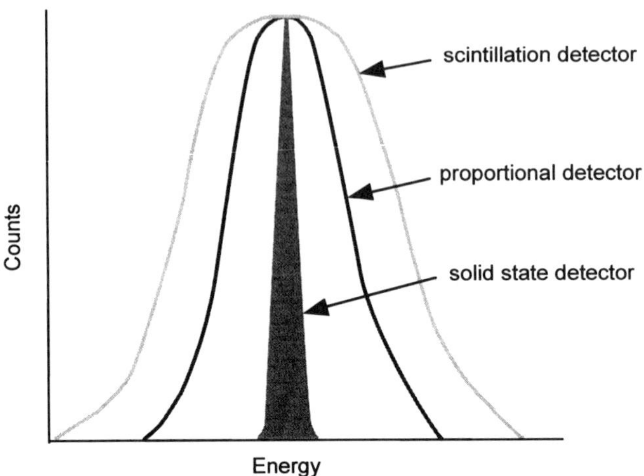

FIG. 51. Pulse-height distribution curves for three kinds of detector. (Peak widths exaggerated.)

tor because it has a high electrical resistivity, particularly at low temperatures, since very few electrons have enough energy to cross the energy band gap. This characteristic is important because the only electrons we want to cross the energy band gap are those excited by an x-ray photon. As we discuss in more detail later, thermally excited electrons just produce noise in the diffraction pattern.

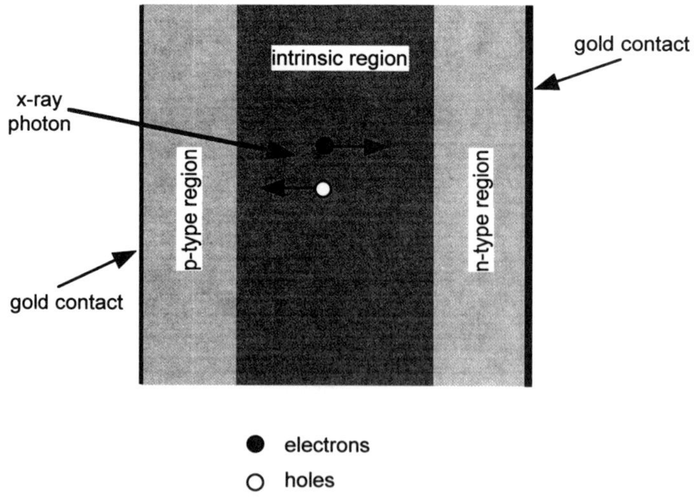

FIG. 52. Solid-state detector.

Intrinsic silicon is very difficult to make commercially and usually contains impurities (mostly boron) that make it act as a *p*-type semiconductor (i.e., there are excess holes). To compensate for these unintentionally present impurities, we add lithium (we say that the Li is "drifted" in) either by thermal diffusion under an applied potential or by ion implantation. On one side of the silicon crystal the lithium concentration is high, and on the other side it is low. The lithium exists in the crystal as Li^+, so, as indicated by the following reaction, free electrons are produced:

$$\mathrm{Li} \rightarrow \mathrm{Li}^+ + \mathrm{e}^-$$

One side of the crystal (where the concentration of Li^+ is high) becomes *n*-type (because there are now excess free electrons). In the middle there is a region where the crystal is intrinsic because the free electrons produced by the reaction have recombined with the excess holes that were already there. The other side of the crystal is *p*-type because of the impurities present in the original crystal. The crystal is called a lithium-drifted silicon crystal and denoted Si(Li). Very thin films of gold are deposited onto the front and back faces of the crystal for the electrical contacts.

When x-rays interact with the silicon crystal they excite electrons from the valence band into the conduction band, creating an electron–hole pair (the band gap in silicon is 1.1 eV). When a reverse-bias potential is applied to the crystal (i.e., a negative charge is applied to the *p*-type region and a positive charge to the *n*-type region), the electrons and holes are separated and a charge pulse of electrons can be measured. The number of electrons or holes created is directly proportional to the energy of the incoming x-ray. Creation of an electron–hole pair in silicon at 77 K, the operating temperature of the detector, requires an average energy of 3.8 eV. It is important to realize that this energy is not the energy of the band gap. Silicon has an indirect band gap (a topic beyond the scope of this book), and transitions from the valence band to the conduction band involve the participation of a phonon (a quantum unit of lattice vibrational energy). Direct transitions (without the help of a phonon) are not permissible below about 3.4 eV.

The number of electron–hole pairs (n) created by absorption of one photon is

$$n = \frac{\text{energy of the photon}}{\text{energy required to create one electron–hole pair}} \tag{29}$$

The absorption of a Cu Kα photon, with energy 8.04 keV, should therefore create 8040/3.8 = 2116 pairs. However, successive photons create slightly different numbers of electron–hole pairs. For example, the first photon may produce 2110 electron–hole pairs, the second 2121 pairs, the third 2126 pairs, and so on. This difference leads to a variation in the size of the output pulse and is the reason for the peak width in Fig. 51. Even so, solid-state detectors produce much narrower peaks (have better energy resolution) than either proportional detectors or scintillation detectors (Fig. 51).

The Si(Li) solid-state detector must be kept at liquid nitrogen temperature (77 K or –196°C) at all times (even when not in use). We have to cool the detector for two reasons: (1) minimize thermal excitation of electrons; at room temperature, thermal energy is sufficient to produce numerous electron–hole pairs, giving a noise level that would swamp the x-ray signals we want to detect; (2) minimize thermal diffusion of lithium, which would upset the distribution obtained during detector fabrication.

Many solid-state detectors are cooled with a Dewar filled with liquid nitrogen connected to the Si(Li) detector via a cold finger. In newer solid-state detectors, however, cooling is achieved by Peltier thermoelectric cooling. Such detectors are small and lightweight, and there is no need for the bulky liquid nitrogen Dewar.

The solid-state detector is extremely efficient, almost 100% over a large range of energies, where efficiency is defined as the ratio of the number of pulses produced to the number of photons striking the detector.

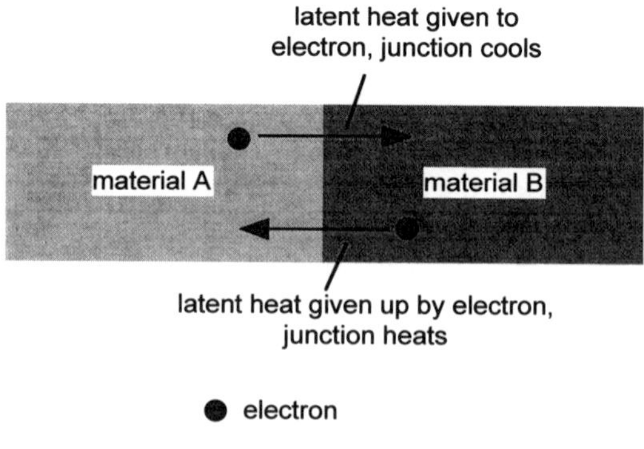

FIG. 53. Peltier effect.

Peltier Effect

When a current flows through a junction of dissimilar conductors, heat can be absorbed or liberated, depending on the current polarity. This phenomenon is known as the *Peltier effect*. Semiconductors are particularly suitable for making devices that utilize the Peltier effect because of their high efficiencies. For semiconductor devices the junction can be a *p–n* junction (i.e., a junction formed between a *p*-type semiconductor and an *n*-type semiconductor) or a metal–semiconductor junction. If electron transfer across the junction results in the electron's absorbing energy, then a cooling effect is observed (Fig. 53). This effect is required for refrigeration applications.

Silicon is not the only semiconductor used for a solid-state x-ray detector. Germanium single-crystal detectors are widely used for γ-rays and high-energy (>20 keV) x-rays because Ge is heavier than Si and therefore is a better absorber of the more energetic radiation. It is easier to produce high-purity single crystals of Ge than it is silicon, so Li drifting is not required. Another advantage of Ge detectors is that they have better energy resolution and a higher signal-to-noise ratio than Si detectors, because the energy to create an electron–hole pair is only about 2.9 eV compared with about 3.8 eV in silicon. The resolution R is inversely proportional to $\sqrt{n}$:

$$R \propto \frac{1}{\sqrt{n}} \tag{30}$$

where n is the number of electron–hole pairs. Using Eq. (29), we obtain for the absorption of a Cu Kα photon by a Ge crystal 8040/2.9 = 2772 electron–hole pairs (remember for Si we were getting about 2116 electron–hole pairs per incoming x-ray photon). Typical best values of resolution for a solid-state detector are 145 eV for a high-purity Ge detector and 165 eV for a Si(Li) detector. Since the resolution is the *smallest* measurable difference in a quantity, in this case energy (the resolution, or resolving power, of an electron microscope is much smaller than that of a visible light microscope—one reason why electron microscopes are essential in the study of materials), the smaller the resolution the better the detector.

Of course, there are some disadvantages with using Ge detectors. Silicon solid-state detectors are slightly more efficient than Ge detectors from 1 to 20 keV. The useful x-ray energies for x-ray diffraction studies

Escape Peaks

Not all the x-ray energy incident upon the detector is transformed to electron–hole pairs. Some of the energy may produce fluorescent radiation that escapes from the intrinsic region of the detector, carrying with it energy that would normally be absorbed. An escape peak is then formed. More escape peaks occur in Ge because we can produce both the Ge Kα (9.89 keV) and Lα (1.19 keV) characteristic x-rays in the detector. X-ray photons having an energy a little higher than the critical ionization energy of the detector material are the most efficient at producing fluorescence. For example, the most efficient x-ray to fluoresce Si Kα x-rays is the P Kα. When the x-ray energy is much larger than the critical ionization energy, x-rays are not absorbed by the detector.

fall within this energy range (for example, Cu Kα x-rays have an energy of 8.04 keV). Si(Li) solid-state detectors are easier to manufacture and are more reliable than Ge detectors. Also, there are more interferences from escape peaks. Germanium solid-state detectors are not used on conventional x-ray diffractometers.

The next generation of solid-state detectors may well use *p–i–n* diodes based on compound semiconductors (a semiconductor consisting of two or more elements). These devices can be made very thin, and it is possible to control the energy of the band gap. If the semiconductor has a large band gap (i.e., the top of the valence band and the bottom of the conduction band are separated by a large amount of energy), then thermal excitation of electrons becomes less probable and only an incident x-ray has enough energy to produce a significant number of electron–hole pairs. Use of these so-called wide-band-gap semiconductors makes it possible to eliminate cooling the detector. The best energy resolution for a detector using a wide-band-gap semiconductor is approaching that of the elemental solid-state detectors.

3.3. EXAMINATION OF A STANDARD X-RAY DIFFRACTION PATTERN

An example of a typical x-ray diffraction pattern is shown in Fig. 54a. The pattern is actually from aluminum, but the general features of the pattern are independent of the specimen. The pattern consists of a series of peaks. The peak intensity is plotted on the ordinate (*y* axis) and the measured diffraction angle, 2θ, along the abscissa (*x* axis). As in Sec. 2.7,

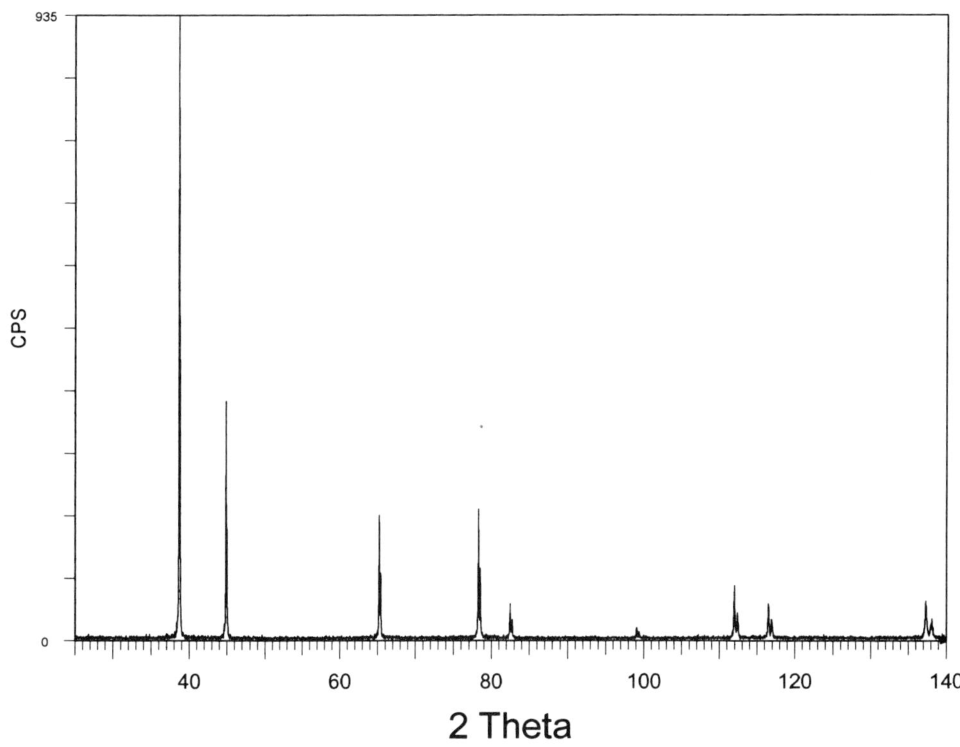

FIG. 54. (a) X-ray diffraction pattern of aluminum.

the peaks are also called reflections. (When recorded in a camera, they were always called lines.) Each peak, or reflection, in the diffraction pattern corresponds to x-rays diffracted from a specific set of planes in the specimen, and these peaks are of different heights (intensities). The intensity is proportional to the number of x-ray photons of a particular energy that have been counted by the detector for each angle 2θ. The intensity is usually expressed in arbitrary units because it is difficult to measure the absolute intensity. Fortunately, we are generally not interested in the absolute intensity of the peak but rather the relative intensities of the peaks and the relative differences in their integrated intensity (the area under the peak). The intensities of the reflections depend on several factors, including structure factor (Sec. 2.8), incident intensity, slit width, and values of current and voltage used in the x-ray source.

The as-recorded x-ray diffraction patterns generally have a background. The background is usually subtracted and the peaks smoothened. We have done this in Fig. 54a. Further, since the most intense peak may have a much higher intensity than the other peaks in the diffraction

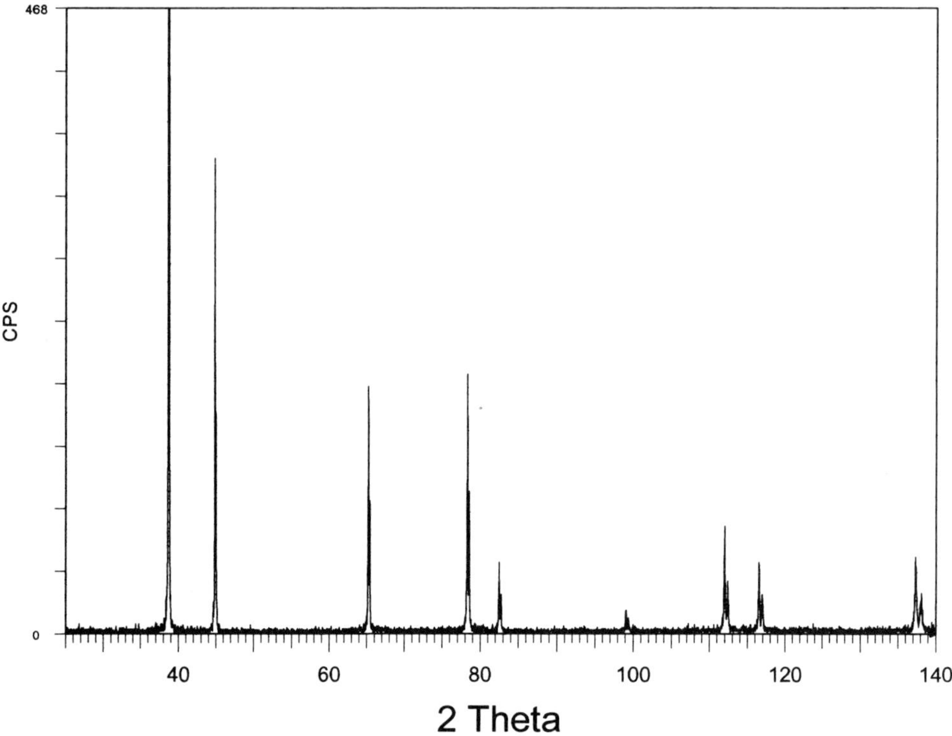

FIG. 54. (b) X-ray diffraction pattern of aluminum where the relative peak intensities have been increased by cutting off the peak height of the most intense peak.

pattern, the intensity scale is often adjusted so as to have a reasonable intensity for all peaks. We have done this in Fig. 54b. The 111 reflection (the first one) is off-scale but the other reflections are now more clearly visible. Hence, you should use these patterns not to evaluate the intensities but to locate peak positions. However, if the intensities of the reflections are required, as in Experimental Modules 7 and 8, the x-ray diffraction patterns are presented without changing the intensity scale. The background is, however, subtracted.

The positions of the peaks in an x-ray diffraction pattern depend on the crystal structure (more specifically, the shape and size of the unit cell) of the material (we are ignoring the effects of texturing and assuming that we have an ideal polycrystalline specimen consisting of randomly oriented equiaxed grains), and this is what enables us to determine the structure and lattice parameter of the material. (The positions of the peaks also depend on the wavelength of the x-rays used. In Exercise 2 at the end of Experimental Module 1 we ask you to show how λ affects the positions of the peaks in the x-ray diffraction pattern of copper.)

As the symmetry of the crystal structure *decreases,* there is an *increase* in the number of peaks. For example, diffraction patterns from materials with cubic structures have few peaks. On the other hand, materials with hexagonal (and other less symmetric) structures show many more peaks in their diffraction patterns. (Compare your results from Experimental Modules 1 and 2 to confirm this observation.)

Diffraction patterns from cubic materials can usually be distinguished at a glance from those of noncubic (less symmetric) materials. Figure 55 shows the calculated diffraction patterns (for a = 0.4 nm and Cu Kα_1 radiation) of the four cubic crystal structures (simple cubic, bcc, fcc, and diamond cubic). You can see from Fig. 55 that in simple cubic and bcc structures, the peaks are approximately equally spaced. In the diffraction pattern from the fcc structure, the peaks appear alternatively as a pair and a single peak. In the diamond cubic structure, the peaks are alternatively more widely and less widely spaced. The reason for these differences in the diffraction patterns is a result of reflections forbidden by the structure factor (see Sec. 2.8). In Experimental Module 1 you will use the structure factor to index x-ray diffraction patterns from a variety of materials with cubic crystal structures. You also need to be aware of which reflections

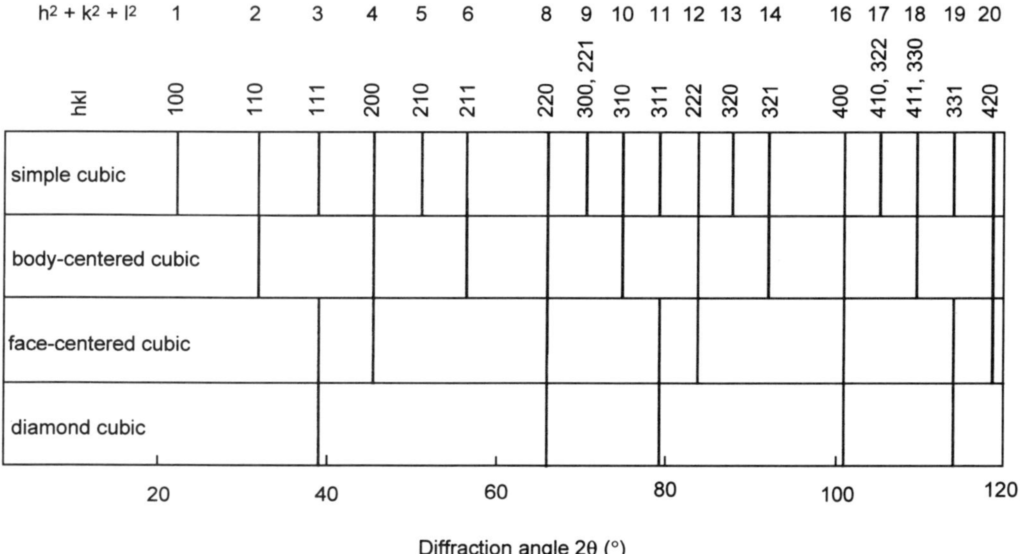

FIG. 55. Comparison of x-ray diffraction patterns from materials with different cubic crystal structures.

are allowed and which are forbidden when you work through some of the other experimental modules.

The intensities of the reflections in a single-phase material provide us with information about the atom positions in the crystal. If you refer to Sec. 2.8 you will see why. In this book we do not go into the exact determination of atom positions from peak intensities. It is beyond what would normally be covered in an undergraduate materials science course, but you will use the relative intensities of the peaks to determine the extent of ordering in the Cu_3Au alloy in Experimental Module 4 or to estimate the relative proportion of phases in Experimental Module 7.

The width of an individual peak, often defined as the full width at half the maximum height (see Fig. 44 for the definition of FWHM), can be used to determine crystallite size and the presence of lattice distortions (strain) in the crystal. You will use peak broadening in Experimental Module 6. (This is somewhat similar to the fine structure in the electron diffraction patterns obtained in a transmission electron microscope.)

For low values of 2θ each reflection appears as a single sharp peak. For larger values of 2θ (above 80° for the aluminum patterns in Fig. 54) each reflection consists of a pair of peaks, which correspond to diffraction of the $K\alpha_1$ and $K\alpha_2$ wavelengths. Recall from Bragg's law ($\lambda = 2d \sin \theta$) that a peak is produced at a Bragg angle θ and for a specific value of d. Even if d is the same, we have one θ value for $K\alpha_1$ and another for $K\alpha_2$. Also note that the separation between the α_1 and α_2 components at any angle is larger when we expand the 2θ scale and, therefore, we may see α_1 and α_2 separated even at lower values (2θ less than 80°). It is useful to remember that a peak *always* contains the α_1 and α_2 components. Whether we see it resolved or not depends on the 2θ values, the scale on the x axis, and the resolution of the diffractometer. At low values of 2θ the separation of the peaks is quite small, but increases at larger 2θ values. The separation of the Cu $K\alpha_1$ and $K\alpha_2$ peaks increases from 0.05° at 20° (2θ) to 1.08° at 150° (2θ). As you will see in Experimental Module 3, the additional peaks (due to α_1 and α_2 resolution) in the x-ray diffraction pattern can be extremely useful.

One parameter that you may very well wish to vary is the length of time it takes to obtain the diffraction pattern. It depends on two factors:

1. The range of 2θ values you want to incorporate into the pattern
2. The length of time you want to spend collecting data at each 2θ value

In most of the experimental modules we give you the suitable range of 2θ values, but for an unknown specimen you may need to cover the full range of angles in order to avoid missing any of the important reflections. The longer you spend collecting data at each angle the better your counting statistics will be—you will have a higher signal-to-noise ratio and better resolution. On the other hand, the amount of time (and money) you have available will provide a practical limit to the length of each scan. The time to record each x-ray diffraction pattern in Part II was between 15 and 60 min.

On modern instruments peak searches are performed on a computer. The computer computes the 2θ angles, the FWHM, and integrated intensities for each peak. In addition, the peak shape can be defined, and overlapping peaks deconvoluted. Most commercial software allows the user to compare standard patterns (from the JCPDS–ICDD data base; see Sec. 3.4) with experimentally observed patterns, allowing rapid matching of patterns and material identification. Certain software packages also allow you to

- Determine lattice strain
- Calculate crystallite size
- Refine calculation of lattice parameter(s) (Rietveld method)
- Calculate diffraction patterns, and other operations

There are continuous advances in the software accompanying x-ray diffractometers, and all these advances help to make life easier.

It is essential that the diffractometer be aligned and calibrated properly. Failure to do so degrades instrument performance, leading to a loss of intensity and resolution, increased background, incorrect profile shapes, and various errors that cannot be readily diagnosed. Procedures and devices for this purpose are often provided by the manufacturer. As a new user of an x-ray diffractometer it is unlikely that you will be required to perform any of the alignment and calibration procedures, but it is important that someone has already done so.

3.4. SOURCES OF INFORMATION

One of the most useful sources of information for crystal structure data is the Powder Diffraction File (PDF). The PDF is a collection of single-phase x-ray powder diffraction patterns in the form of tables of interplanar spacings (d) and corresponding relative intensities. The files also contain other information, such as some physical and crystallographic properties of the material. The origin of the PDF goes back to 1936 when Hanawalt and others started collecting about 1000 standard x-ray diffraction patterns as a repository. Since the number of standard patterns started growing, this effort was taken over in 1941 by several societies, including the American Society for Testing and Materials (ASTM). In 1969 the organization was constituted as a Pennsylvania nonprofit corporation under the title of The Joint Committee on Powder Diffraction Standards (JCPDS). In 1978, the organization was renamed the International Centre for Diffraction Data (ICDD). The acronym JCPDS is still widely used to refer to both the organization and the patterns. In 1976 the PDF comprised about 26,000 diffraction patterns; there are now about 80,000 patterns. The data base is continually growing as new materials are synthesized and their crystal structures determined. The substances included in the PDF are elements, alloys, inorganic compounds, minerals, organic compounds, and organometallic compounds. The PDF is also continually updated, and incorrect patterns are deleted and replaced.

In the early days the full details of each pattern [*d* spacings, relative intensities, Miller indices of planes, optical information, density, number of atoms (molecules) in the unit cell, etc.] were printed in the form of 3″ × 5″ file cards. Later, these cards were printed (three per page) in books (the PDF now consists of 47 sets of cards), and now all the data are available on CD–ROM. In either form these tabulations are invaluable to anyone doing materials characterization using x-ray diffraction (and, by the way, when you do electron diffraction in the transmission electron microscope). Most modern x-ray diffractometers have these files on computer, so they can be easily accessed and printed. People still use the term "cards" to indicate the standard x-ray diffraction pattern, so we use this term in the experimental modules.

The contact information for the International Centre for Diffraction Data is

International Centre for Diffraction Data
Newtown Square Corporate Campus
12 Campus Boulevard
Newtown Square, Pennsylvania 19073-3273
Phone 610-325-9810; Fax 610-325-9823,
Internet INFORMATION@ICDD.COM

Figure 56 shows a typical data "card" obtained from a computer version of the PDF. Most of the information for these cards was generated with a powder camera and Cu Kα radiation in earlier years, but a diffractometer is now used. However, with modern computer facilities you can convert or transform the data to other radiations quite easily. In Table 9 we explain the features of the card (refer to the Fig. 57 for the meanings of the numbers in column 1).

TABLE 9. Details of the Data Available on the JCPDS–ICDD Files

Number	Feature
1	File number. Each card in the PDF has its own unique number. The first number refers to the series. The last four digits (after the dash) represent the actual card number.
2	The wavelength of the radiation used to obtain the diffraction pattern.
3	The quality of the data. In the computer-generated files the quality is given as high, indexed, not indexed, and questionable. In the cards the following symbols are used: * (high quality), i (peaks indexed, intensities fairly reliable), c (calculated pattern), and o (low reliability).
4	The name of the material and its chemical formula. "Syn." means the material was not obtained from a natural source but was synthesized.
5	The diffraction pattern information. The data given are the interplanar spacing, *d* (in Å) (the computer can readily convert the *d* values to the diffraction angle 2θ), *I*, or I/I_1, the relative intensity of the reflections expressed as percentages of the strongest peak in the pattern, and *hkl* are their Miller indices.
6	The lattice and space group (S.G.) of the material.
7	Lattice parameters (*a, b, c*). Note that the lattice parameters are given in Å (1 Å = 0.1 nm = 10^{-10} m).
8	Lattice parameters (α, β, γ) and number of molecules (atoms in the case of pure metals) per unit cell.
9	Molecular weight, volume, and experimental (D_x) and measured (D_m) density in g/cm^3. I/I_{cor} is the ratio of the peak height of the strongest reflection of the specimen to the strongest reflection of corundum (α-Al_2O_3).
10	How the sample was prepared/obtained. Optical and other properties of the material are also included here, although this information may not be complete for each material in the PDF.
11	The reference to the original source of the diffraction pattern information. CAS is the Chemical Abstract Series number.
12	The type of radiation used and its wavelength. The SS/FOM number is the Smith Schneider figure of merit. It gives the total number of reflections in the pattern and provides a measure of data reliability.
13	Other experimental information, including the type of instrument used.

These cards can be retrieved with the help of a computer when you know the card number or the name of the material or the chemical constituents. In the experimental modules you can use the PDF in two ways. First, you can use it to check your indexing. If the diffraction pattern you obtained is from a known material (e.g., silicon), then, after you have indexed the reflections and calculated the lattice parameter, you can check your results against the relevant card number (e.g., for silicon the card number is 27-1402). The card numbers for the materials referred to in the experimental modules are listed in Appendix 9.

In Experimental Module 8 we show you how to use the PDF to identify an unknown material. This method is known as the Hanawalt method and uses the PDF Search Manual (another invaluable resource from ICDD). In the Search Manual the positions of eight peaks (in terms of

Pattern : 27-1402 **Radiation** = 1.540560 **Quality :** High

Si

Silicon, syn [NR] / Silicon

Lattice : Cubic

S.G. : Fd3m (227)

Mol. weight = 28.09

Volume [CD] = 160.18

Dx = 2.329

I/Icor = 4.70

a = 5.43088	**Alpha** = 90.00	
b = 5.43088	**Beta** = 90.00	
c = 5.43088	**Gamma** = 90.00	
a/b = 1.00000	**Z** = 8	
c/b = 1.00000		

d (Å)	I	h	k	l
3.1355	100	1	1	1
1.9201	55	2	2	0
1.6375	30	3	1	1
1.3577	6	4	0	0
1.2459	11	3	3	1
1.1086	12	4	2	2
1.0452	6	5	1	1
0.9600	3	4	4	0
0.9180	7	5	3	1
0.8587	8	6	2	0
0.8282	3	5	3	3

TEMP. OF DATA COLLECTION : Pattern at 25(1) C.
SAMPLE SOURCE OR LOCALITY : This sample is NBS Standard Reference Material No. 640.
GENERAL COMMENTS : Reflections calculated from precision measurement of a0.
GENERAL COMMENTS : a0 uncorrected for refraction.
ADDITIONAL PATTERN : To replace 5-565 and 26-1481.
COLOR : Gray

*Natl. Bur. Stand. (U.S.) Monogr. 25, volume 13, page 35, (1976) primary reference :

Radiation : CuKa1	***Filter :*** Monochromator crystal
Lambda : 1.54060	***d-sp :*** Diffractometer
SS/FOM : F11=408(0.0021,1)	***Internal standard :*** W

FIG. 56. JCPDS–ICDD card for silicon.

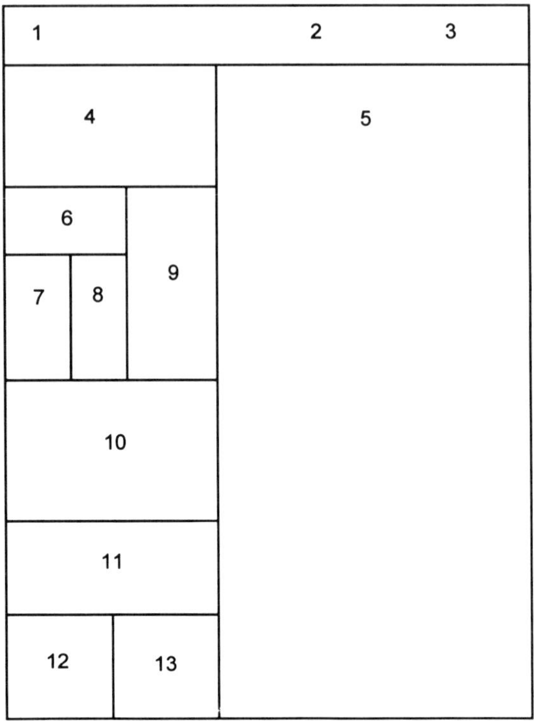

FIG. 57. Schematic of a JCPDS–ICDD card.

their *d* values) from the diffraction pattern are listed in order of decreasing intensity. The subscript under each *d* spacing represents the intensity of the peak rounded off to the nearest 10. For example, 5 represents 50% and 3 represents 30%. The highest intensity, namely 100%, should be represented by 10 but in the listing it is represented by *x*. The *d* value of the strongest peak determines the group (there are 40 groups) into which the material is placed; the *d* value of the next strongest peak specifies the subgroup, i.e., the location of the material within the group. The next strongest six peak positions can be used to confirm the identification. The groups are divided into ranges of *d* (in Å), as shown in Table 10.

To identify an unknown specimen, start by matching the strongest peak, then the next strongest, and so on. If you can match the eight peak positions and intensities in the PDF with the eight strongest peaks and intensities in your experimentally determined diffraction pattern, that is usually more than sufficient to identify the unknown specimen. The details of how to do this are presented in Experimental Module 8. The International Centre for Diffraction Data also publishes an Alphabetical

TABLE 10. Groups of *d* Spacings (in Å) in the PDF Search Manual (Hanawalt Method)

999.99–8.00 (–0.20)	3.31–3.25 (±0.02)	2.64–2.58 (±0.01)
7.99–7.00 (±0.10)	3.24–3.20 (±0.01)	2.57–2.51 (±0.01)
6.99–6.00 (±0.10)	3.19–3.15 (±0.01)	2.50–2.44 (±0.01)
5.99–5.50 (±0.05)	3.14–3.10 (±0.01)	2.43–2.37 (±0.01)
5.49–5.00 (±0.05)	3.09–3.05 (±0.01)	2.36–2.30 (±0.01)
4.99–4.60 (±0.04)	3.04–3.00 (±0.01)	2.29–2.23 (±0.01)
4.59–4.30 (±0.03)	2.99–2.95 (±0.01)	2.22–2.16 (±0.01)
4.29–4.10 (±0.03)	2.94–2.90 (±0.01)	2.15–2.09 (±0.01)
4.09–3.90 (±0.02)	2.89–2.85 (±0.01)	2.08–2.02 (±0.01)
3.89–3.75 (±0.02)	2.84–2.80 (±0.01)	2.01–1.86 (±0.01)
3.74–3.60 (±0.02)	2.79–2.75 (±0.01)	1.85–1.68 (±0.01)
3.59–3.50 (±0.02)	2.74–2.70 (±0.01)	1.67–1.38 (±0.01)
3.49–3.40 (±0.02)	2.69–2.65 (±0.01)	1.37–1.00 (±0.01)
3.39–3.32 (±0.02)		

Index that lists all the compounds in alphabetical order. The index can be used when you know what the material is but do not know the positions of the major reflections.

Other useful sources of information about crystal structure data are

1. P. Villars and L. D. Calvert, *Pearson's Handbook of Crystallographic Data for Intermetallic Phases* (in four volumes), ASM International, Materials Park, OH (1991); includes information on pure metals, semiconductors, and ceramic compounds, in addition to intermetallic phases.
2. W. B. Pearson, *A Handbook of Lattice Spacings and Structures of Metals and Alloys*, Vol. 1 (1958) and Vol. 2 (1967), Pergamon Press, Oxford, UK; lists lattice spacings of alloys and compounds as a function of solute content.

3.5. X-RAY SAFETY

It is important to be aware of the potential hazards associated with the use of x-ray diffraction apparatus. With modern x-ray equipment these hazards have been reduced to a minimum, and with responsible use of the equipment you should not be in any danger. However, x-rays are extremely dangerous and must be treated with caution and respect.

X-rays are damaging because they penetrate the human body and break up molecules of DNA in our cells. Sometimes the DNA molecules repair themselves and there is no lasting effect of the radiation. Sometimes the molecules are not repaired and the cells die. If this process occurs on the surface of the skin, then the skin will be sloughed, much like a bad case of sunburn. However, if after the DNA molecules have been broken they rejoin in the wrong way, then a mutant cell is produced and this can lead to cancer.

Most students who use x-ray diffraction and other x-ray instruments experience no significantly greater exposure to radiation than a normal member of the public. In the United States the average exposure to radiation is about 360 mrem/year. The largest contribution to this dose is indoor radon inhalation, which averages about 200 mrem/year in the United States. Although we cannot avoid exposure to radiation, we want to keep our exposure *as low as reasonably achievable (alara).*

How can you keep your exposure *alara*? Three factors limit your total exposure:

1. Distance
2. Time
3. Shielding

The further you are from the source of radiation, the less time you spend in the vicinity of a radiation source, and the greater the amount of shielding between you and the x-ray source the less your exposure will be. Because x-rays are so energetic, they are very penetrating; hence, the most effective shielding material has a high mass absorption coefficient, i.e., a large atomic number (Z). The most common shielding material is lead ($Z = 82$). However, although the intensity of radiation can be reduced ("attenuated" is the term used by x-ray safety people) by shielding, it cannot be completely eliminated.

Most x-ray diffractometers have a safety screen (or door) that encloses the instrument. The screen is opened before loading your specimen. While the screen is open, the x-ray source cannot be operated. Never tamper with this safety mechanism. After your specimen is loaded into the diffractometer, the screen is closed and the x-ray source can be energized. When you want to change specimens, turn off the x-ray source and lift

Units of Radiation Dose

The rem is a unit of radiation dose, known as the biological equivalent dose, and is the amount of energy absorbed by the body multiplied by a factor related to how effectively the radiation is absorbed by the body. The rem has been superseded by the SI unit the sievert (Sv): 1 Sv = 100 rem.

the screen. There are no residual x-rays in the chamber after the source has been turned off. X-rays are electromagnetic radiation like visible light—as soon as you turn off the light the room is dark!

Brief, occasional exposures to low levels of x-rays are not cumulative. Most users of x-ray diffraction instruments do not show any significant exposure effects or higher levels of radiation dose than does the general public. If you have any doubts about the level of radiation in your laboratory, then a simple blood test can determine the amount of your exposure. Frequent users of the x-ray equipment may wear film badges that can also be used to determine the level of exposure.

You will probably see warning signs posted near the x-ray diffractometer and on the door of the laboratory. Also, notices will be posted describing various emergency procedures and regulations. It is worth reading these notices the first time you use the diffractometer. There will probably be a log book near the instrument filled in by the various users. It is important that you complete this log after your session on the instrument and if you have any problems be sure to point these out to the laboratory technician or the radiation safety officer at your institution. If you use x-ray diffraction a lot, particularly for graduate work, you will probably be required to take a seminar on radiation safety.

3.6. INTRODUCTION TO THE EXPERIMENTAL MODULES

Part II consists of eight experimental modules that represent the types of studies ideally performed with an x-ray diffractometer. The emphasis here is on polycrystalline materials (mainly powders) and not single crystals. X-ray diffraction can be used to obtain structural information about single crystals, but this requires a slightly different experimental setup that is not available in many undergraduate laboratories. Powder x-ray diffraction is used in a wide variety of disciplines, and thousands of papers have been published on the structures of all classes of materials: metals, ceramics, semiconductors, and polymers. Studies well suited to powder x-ray diffraction include

- Identification of crystalline phases
- Qualitative and quantitative analysis of mixtures and minor constituents
- Distinction between crystalline and amorphous states
- Following solid-state reactions
- Identification of solid solutions
- Identification of polymorphism

- Phase diagram determination
- Lattice parameter measurements
- Detection of preferred orientation
- Microstructural characterization
- *In situ* temperature and pressure studies

The experimental modules are self-contained experiments that show you how to obtain specific information from an x-ray diffraction pattern. The topics covered by the modules are as follows:

Module 1 Crystal Structure Determination. I: Cubic Structures
Module 2 Crystal Structure Determination. II: Hexagonal Structures
Module 3 Precise Lattice Parameter Measurements
Module 4 Phase Diagram Determination
Module 5 Detection of Long-Range Ordering
Module 6 Determination of Crystallite Size and Lattice Strain
Module 7 Quantitative Analysis of Powder Mixtures
Module 8 Identification of an Unknown Specimen

These modules describe the major applications of x-ray diffraction in materials science. By completing all the modules, you will be fully conversant with the full potential of x-ray diffraction of polycrystalline materials. For example, you will be able to index diffraction patterns from materials with cubic and hexagonal structures. Diffraction patterns from materials with other crystal structures can also be indexed by using the appropriate plane spacing equations (see Appendix 1). However, materials with structures less symmetric than cubic produce complex patterns containing many reflections. You will also be able to determine the phase proportions in a multiphase mixture and the fine structure of the materials (e.g., state of ordering, crystallite size, strain). You will also be able to identify an unknown material by using x-ray diffraction techniques, starting from fundamental principles.

As mentioned, however, you will not be doing a complete structure determination, since methods of determining atom positions are not covered in these modules. Neither measurements of stress in the specimens nor texturing will be covered; these topics are highly specialized and require special adaptations or arrangements of the diffractometer.

At the end of each module are exercises to give you a chance to apply some of the knowledge you acquired by working through the module. All the information you need to answer the exercises is given either in Part I or in the module itself.

Part II

Experimental Modules

EXPERIMENTAL MODULE 1

Crystal Structure Determination. I: Cubic Structures

OBJECTIVE OF THE EXPERIMENT

To index the x-ray diffraction pattern, identify the Bravais lattice, and calculate the lattice parameters of some common materials with a cubic structure.

MATERIALS REQUIRED

Fine powders (–325 mesh, <45 μm in size) of chromium (Cr), copper (Cu), iron (Fe), nickel (Ni), silicon (Si), and titanium nitride (TiN).

BACKGROUND AND THEORY

A knowledge of crystal structure is a prerequisite to understand phenomena such as plastic deformation, alloy formation, and phase transformations. Recall that the size and shape of the unit cell determine the angular positions of the diffraction peaks, and the arrangement of the atoms within the unit cell determines the relative intensities of the peaks. Thus, it is possible to calculate the size and shape of the unit cell from the angular positions of the peaks and the atom positions in the unit cell (though not in any direct manner) from the intensities of the diffraction peaks.

Complete determination of an unknown crystal structure consists of three steps:

1. Calculation of the size and shape of the unit cell from the angular positions of the diffraction peaks
2. Computation of the number of atoms per unit cell from the size and shape of the unit cell, the chemical composition of the specimen, and its measured density
3. Deduction of the atom positions within the unit cell from the relative intensities of the diffraction peaks

In this experimental module, we only do step 1 (assuming that the shape is cubic).

Indexing the Pattern

"Indexing the pattern" involves assigning the correct Miller indices to each peak in the diffraction pattern. These peaks are also referred to as reflections, and as in Part I, we use both terms interchangeably. It is important to remember that correct indexing is done only when *all* the peaks in the diffraction pattern are accounted for and no peaks expected for the structure are missing from the diffraction pattern.

A typical example of indexing diffraction patterns obtained from materials with a cubic structure is presented here. The procedure is the same for a metal, a semiconductor, or a ceramic.

The interplanar spacing d, the distance between adjacent planes in the set (hkl) of a material with a cubic structure and lattice parameter a, may be obtained from the equation

$$\frac{1}{d^2} = \frac{h^2 + k^2 + l^2}{a^2} \tag{1.1}$$

Combining Bragg's law ($\lambda = 2d \sin \theta$) with Eq. (1.1) yields

$$\frac{1}{d^2} = \frac{h^2 + k^2 + l^2}{a^2} = \frac{4 \sin^2 \theta}{\lambda^2} \tag{1.2}$$

Rearranging gives

$$\sin^2 \theta = \left(\frac{\lambda^2}{4a^2}\right)(h^2 + k^2 + l^2) \tag{1.3}$$

Realizing that $\lambda^2 / 4a^2$ is a constant for any one pattern, we note that $\sin^2 \theta$ is proportional to $h^2 + k^2 + l^2$; i.e., as θ increases, planes with higher Miller indices will diffract. Writing Eq. (1.3) for two different planes and dividing, we get

$$\frac{\sin^2\theta_1}{\sin^2\theta_2} = \frac{h_1^2 + k_1^2 + l_1^2}{h_2^2 + k_2^2 + l_2^2} \tag{1.4}$$

In the cubic system, the first reflection in the diffraction pattern is due to diffraction from planes with the Miller indices (100) for primitive cubic, (110) for body-centered cubic, and (111) for face-centered cubic lattices (planes with the highest planar density in each case), so $h^2 + k^2 + l^2 = 1$, 2, or 3, respectively. Since the ratio of $\sin^2\theta$ values scales with the ratio of $h^2 + k^2 + l^2$ values for different planes, and since h, k, and l are always integers, the $h^2 + k^2 + l^2$ values can be obtained by dividing the $\sin^2\theta$ values of different reflections with the minimum one (i.e., $\sin^2\theta$ of the first reflection) and multiplying the ratios thus obtained by an appropriate integer. Therefore, the $\sin^2\theta$ values calculated for all peaks in the diffraction pattern are divided by the smallest value (first reflection). These ratios, when multiplied by 2 or 3, yield integers (if they are not already integers). These subsequent integers represent the $h^2 + k^2 + l^2$ values; therefore, the hkl values can be easily identified from the quadratic forms listed in Appendix 2. You may also index the diffraction pattern using the ratio of $1/d^2$ values instead of the $\sin^2\theta$ values. Since θ is a directly measurable quantity we have chosen to use the $\sin^2\theta$ values.

Identification of the Bravais Lattice

The Bravais lattice can be identified by noting the systematic presence (or absence) of reflections in the diffraction pattern (refer to Sec. 2.8 in Part I for more details). Table 1.1 summarizes the selection rules (or extinction conditions as they are also known) for cubic lattices. According to these selection rules, the $h^2 + k^2 + l^2$ values for the different cubic lattices follow the sequence

Primitive 1, 2, 3, 4, 5, 6, 8, 9, 10, 11, 12, 13, 14, 16, . . .
Body-centered 2, 4, 6, 8, 10, 12, 14, 16, . . .
Face-centered 3, 4, 8, 11, 12, 16, 19, 20, 24, 27, 32, . . .

TABLE 1.1. Selection Rules for the Presence or Absence of Reflections

Bravais Lattice	Reflections Present for	Reflections Absent for
Primitive	all	none
Body-centered	$h + k + l$ even	$h + k + l$ odd
Face-centered	h, k, and l unmixed (all even or all odd)	h, k, and l mixed

The diamond cubic structure (adopted by diamond and the elemental semiconductors silicon and germanium) can be considered as two interpenetrating face-centered cubic lattices (see Sec. 2.4.2 in Part I). Thus, in a diamond cubic structure only reflections with h, k, and l even or odd (unmixed) are possibly present. However, due to destructive interference from some of the atoms in certain planes, some additional reflections will be absent (even if h, k, and l are odd or even). (We went through the structure factor calculation for the diamond cubic structure in Sec. 2.8 in Part I.) Thus, the $h^2 + k^2 + l^2$ sequence for the diamond cubic structure is

$$3, 8, 11, 16, 19, 24, 27, 32, \ldots$$

The Bravais lattice and the crystal structure can be thus unambiguously identified from the sequence of $h^2 + k^2 + l^2$ values in the diffraction pattern.

If the diffraction pattern contains only six peaks and if the ratio of the $\sin^2 \theta$ values is 1, 2, 3, 4, 5, and 6 for these reflections, then the Bravais lattice may be either primitive cubic or body-centered cubic; it is not possible to unambiguously distinguish the two. But remember that the simple cubic structure is not very common and, therefore, in such a situation, you will probably be right if you had indexed the pattern as belonging to a material with the bcc structure. However, if the structure is actually simple cubic and the pattern is wrongly indexed as bcc, the first reflection will be indexed as 110 instead of 100, and consequently the lattice parameter of the unit cell will be $\sqrt{2}$ times larger than the actual value. This confusion can be avoided if the x-ray diffraction pattern has at least seven peaks. In that case, the seventh peak has Miller indices 321 ($h^2 + k^2 + l^2 = 14 = 7 \times 2$) for the body-centered cubic lattice and 220 ($h^2 + k^2 + l^2 = 8$) for the primitive cubic lattice. Note that 7 cannot be expressed as the sum of three squares, and therefore the seventh peak in the diffraction pattern from a primitive cubic lattice will be much more separated from the sixth peak than the earlier sets (as illustrated in Fig. 55 in Part I). A large number of peaks in the diffraction pattern can be obtained if the pattern is recorded with a radiation with a short wavelength. For example, use Mo Kα instead of Cu Kα.

Calculation of the Lattice Parameter

The lattice parameter a can be calculated from

$$\sin^2\theta = \left(\frac{\lambda^2}{4a^2}\right)(h^2 + k^2 + l^2) \tag{1.3}$$

Rearranging gives

$$a^2 = \frac{\lambda^2}{4\sin^2\theta}(h^2 + k^2 + l^2) \tag{1.5}$$

or

$$a = \frac{\lambda}{2\sin\theta}\sqrt{h^2 + k^2 + l^2} \tag{1.6}$$

WORKED EXAMPLE

Let's now do an example of indexing the x-ray diffraction pattern of a material with a cubic structure. The x-ray diffraction pattern of aluminum recorded with Cu Kα radiation is presented in Fig. 1.1. Nine peaks exist in the diffraction pattern, and the 2θ values for these peaks are listed in Table 1.2. (Even though the α_1 and α_2 components are resolved for reflections with 2θ > about 80°, we have listed only the 2θ values corresponding to the α_1 component). The $\sin^2\theta$ values have been calculated, and following the described procedure we have completely indexed the diffraction pattern and determined the lattice parameter of aluminum.

Note that only reflections with unmixed *hkl* values (all odd or all even) are present; and therefore, it can be concluded that the Bravais lattice is

TABLE 1.2. Details of Indexing the X-Ray Diffraction Pattern of Aluminum

Material: Aluminum			Radiation: Cu Kα				$\lambda_{K\alpha 1}$ = 0.154056 nm
Peak #	2θ (°)	$\sin^2\theta$	$\frac{\sin^2\theta}{\sin^2\theta_{min}}$	$\frac{\sin^2\theta}{\sin^2\theta_{min}} \times 3$	$h^2 + k^2 + l^2$	*hkl*	*a* (nm)
1	38.52	0.1088	1.000	3.000	3	111	0.40448
2	44.76	0.1450	1.333	3.999	4	200	0.40457
3	65.14	0.2898	2.664	7.992	8	220	0.40471
4	78.26	0.3983	3.661	10.983	11	311	0.40480
5[a]	82.47	0.4345	3.994	11.982	12	222	0.40480
6[a]	99.11	0.5792	5.324	15.972	16	400	0.40485
7[a]	112.03	0.6876	6.320	18.960	19	331	0.40491
8[a]	116.60	0.7238	6.653	19.958	20	420	0.40491
9[a]	137.47	0.8684	7.982	23.945	24	422	0.40494

[a]The α_1 and α_2 peaks have been resolved for these reflections. But, since we do not need a very high accuracy, we have taken the 2θ values of the α_1 component only.

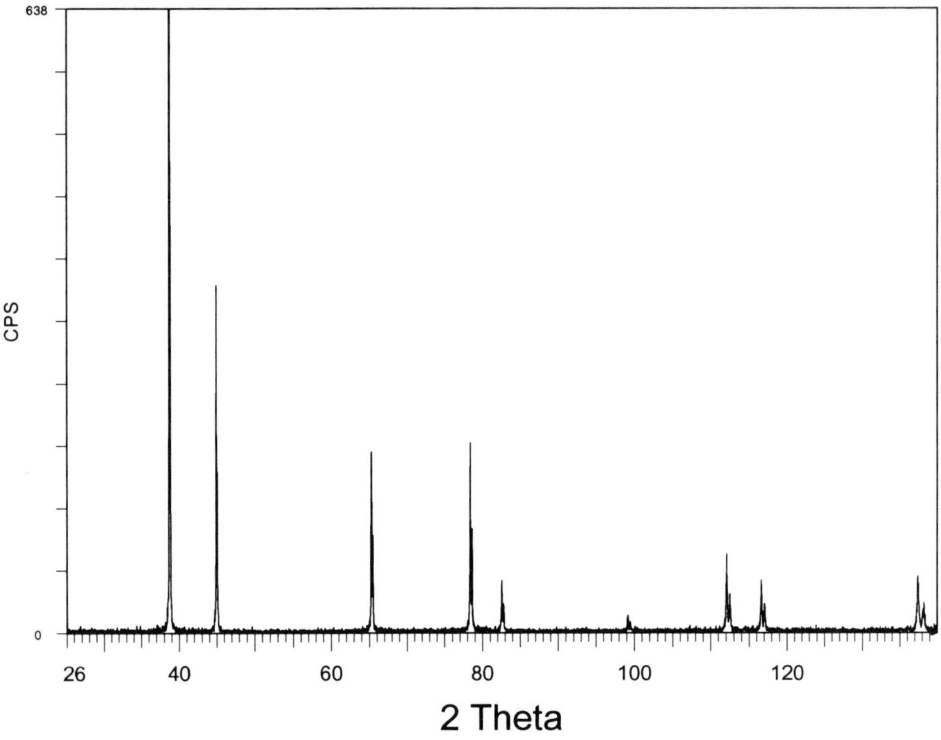

FIG. 1.1. X-ray diffraction pattern of aluminum.

face-centered cubic. The lattice parameter is calculated as 0.4049 nm. Thus, aluminum has an fcc structure with lattice parameter a = 0.4049 nm.

From Table 1.2 you can see that the lattice parameter increases with increasing 2θ values. Since the systematic error in $\sin^2\theta$ decreases as θ increases, you should select the value of a for the highest-angle peak in the diffraction pattern as the most accurate. (See Experimental Module 3 for precise lattice parameter measurements.)

ANALYTICAL METHOD

An analytical approach can also be used to index diffraction patterns from materials with a cubic structure. Look again at Eq. (1.3):

$$\sin^2\theta = \left(\frac{\lambda^2}{4a^2}\right)(h^2 + k^2 + l^2) \tag{1.3}$$

Since $\lambda^2/4a^2$ is constant for any pattern, which we call A, we can write

$$\sin^2\theta = A(h^2 + k^2 + l^2) \tag{1.7}$$

In a cubic system, the possible $h^2 + k^2 + l^2$ values are 1, 2, 3, 4, 5, 6, 8, . . . (even though all may not be present in every type of cubic lattice). The observed $\sin^2\theta$ values for all peaks in the diffraction pattern are therefore divided by the integers 2, 3, 4, 5, 6, . . . to obtain a common quotient, which is the value of A, corresponding to $h^2 + k^2 + l^2 = 1$. We can then calculate the lattice parameter from the value of A by using the relationship

$$A = \frac{\lambda^2}{4a^2} \quad \text{or} \quad a = \frac{\lambda}{2\sqrt{A}} \tag{1.8}$$

Dividing the $\sin^2\theta$ values of the different reflections by A gives the $h^2 + k^2 + l^2$ values, from which the hkl values of each reflection can be obtained by using the listing in Appendix 2.

The aluminum x-ray diffraction pattern indexed earlier is now indexed with the analytical approach, and the results are presented in Table 1.3.

Worked Example

The lowest common $\sin^2\theta$ value in Table 1.3 is 0.0363 (indicated in bold). Note that 0.1086 also appears as a common $\sin^2\theta$ value in some of the columns, but this is approximately 3 × 0.0363. Therefore, you must look for the smallest value of $\sin^2\theta$ among the values. However, realize

TABLE 1.3. Calculation of the Common Quotient

Material: Aluminum		Radiation: Cu Kα					$\lambda_{K\alpha 1}$ = 0.154056 nm	
Peak #	2θ (°)	$\sin^2\theta$	$\frac{\sin^2\theta}{2}$	$\frac{\sin^2\theta}{3}$	$\frac{\sin^2\theta}{4}$	$\frac{\sin^2\theta}{5}$	$\frac{\sin^2\theta}{6}$	$\frac{\sin^2\theta}{8}$
1	38.52	0.1088	0.0544	**0.0363**	0.0272	0.0218	0.0181	0.0136
2	44.76	0.1450	0.0725	0.0483	**0.0363**	0.0290	0.0242	0.0181
3	65.14	0.2898	0.1449	0.0966	0.0725	0.0580	0.0483	**0.0362**
4	78.26	0.3983	0.1992	0.1328	0.0996	0.0797	0.0664	0.0498
5	82.47	0.4345	0.2173	0.1448	0.1086	0.0869	0.0724	0.0543
6	99.11	0.5792	0.2896	0.1931	0.1448	0.1158	0.0965	0.0724
7	112.03	0.6876	0.3438	0.2292	0.1719	0.1375	0.1146	0.0860
8	116.60	0.7238	0.3619	0.2413	0.1810	0.1448	0.1206	0.0905
9	137.47	0.8684	0.4342	0.2895	0.2171	0.1737	0.1447	0.1086

> Note that the common quotient of 0.0363 appears only in the columns of $(\sin^2 \theta)/3$, $(\sin^2 \theta)/4$, and $(\sin^2 \theta)/8$ in Table 1.3. This suggests that the sequence of $h^2 + k^2 + l^2$ values will be in the ratio 3, 4, 8, . . . From the sequence of $h^2 + k^2 + l^2$ values mentioned for different types of cubic lattices, it is clear that the Bravais lattice of aluminum is face-centered cubic, which will also be proved when we divide the $\sin^2 \theta$ values by A. The other Bravais lattices can also be identified in a similar way.

that this value (0.0363) does not (and will not) necessarily appear in each column of the table.

We can now assume that 0.0363 represents the A value in Eq. (1.7) for $h^2 + k^2 + l^2 = 1$. Since Eq. (1.7) suggests that division of the $\sin^2 \theta$ values by A yields the $h^2 + k^2 + l^2$ values from which the hkl values can be obtained (see Appendix 2), the calculations can be tabulated as shown in Table 1.4.

Notice from Table 1.4 that the Miller indices of the reflections again are unmixed (all odd or all even), so the Bravais lattice is face-centered cubic. The lattice parameter is calculated from Eq. (1.8). Since $A = 0.0363$ and $a = \lambda/2\sqrt{A}$, for aluminum

$$a = \frac{0.154056}{2\sqrt{0.0363}} = 0.4049 \text{ nm}$$

As you would expect, the results obtained by this analytical method are the same as those obtained by the earlier method.

TABLE 1.4. Indexing the Diffraction Pattern of Aluminum Using the Analytical Approach

Material: Aluminum		Radiation: Cu Kα		A = 0.0363
Peak #	$\sin^2 \theta$	$\frac{\sin^2 \theta}{A}$	$h^2 + k^2 + l^2$	hkl
1	0.1088	2.997	3	111
2	0.1450	3.995	4	200
3	0.2898	7.983	8	220
4	0.3983	10.972	11	311
5	0.4345	11.970	12	222
6	0.5792	15.956	16	400
7	0.6876	18.942	19	331
8	0.7238	19.939	20	420
9	0.8684	23.923	24	422

In Table 1.4 notice that the value of $(\sin^2 \theta)/A$ deviates more from the expected value at higher diffraction angles than at lower angles. This is so because a small error in the measurement of θ at low angles leads to a large error in the value of sin θ (also in $\sin^2 \theta$). Since the $\sin^2 \theta$ value increases with increasing θ, a small error in the lowest value of $\sin^2 \theta$ gets multiplied several times at higher angles because we are dividing all the $\sin^2 \theta$ values by the smallest $\sin^2 \theta$ value.

EXPERIMENTAL PROCEDURE

Take fine powders (–325 mesh, < 45 μm in size) of chromium (Cr), copper (Cu), iron (Fe), nickel (Ni), silicon (Si), and titanium nitride (TiN). Load each powder separately into the x-ray holder and record the x-ray diffraction patterns, using Cu Kα radiation in the 2θ angular range of 25° to 140°. Using the cursor, locate the maxima of all the peaks in each diffraction pattern and list the 2θ values in increasing order. If the α_1 and α_2 peaks are resolved, you are advised to take the 2θ value of the α_1 peak, even though for greater accuracy you could use both peaks and use the respective wavelengths to calculate the lattice parameters. (In case an x-ray diffractometer is not available, or if you are not able to record the diffraction pattern for some reason, the x-ray diffraction patterns for Cr, Cu, Fe, Ni, Si, and TiN are provided in Figs. 1.2 through 1.7, and the 2θ values of all the reflections are listed in the work tables provided for each different material.) Using the worked example for aluminum, index all the diffraction patterns (find the Miller indices of the planes giving rise to the diffraction peaks). From the systematic absence (or presence) of reflections in each pattern and from Table 1.1, identify the Bravais lattice. Calculate the lattice parameter in each case and tabulate your calculations in the work tables provided.

Note 1: Indexed diffraction patterns (2θ, *hkl*, and intensity values) and the lattice parameters of all standard substances can be found in the Powder Diffraction Files (PDF) published by the JCPDS–International Centre for Diffraction Data (see Sec. 3.4 in Part I for more information about the PDF). These files are available with most computers associated with modern x-ray diffractometers. The "card" numbers for the materials referred to in this book are listed in Appendix 9.

Note 2: Even though diffraction patterns (from different types of cubic structures) have been provided in this module, you may index all of them or only some of them, depending on the time available.

When you finish these calculations, complete Table 1.17.

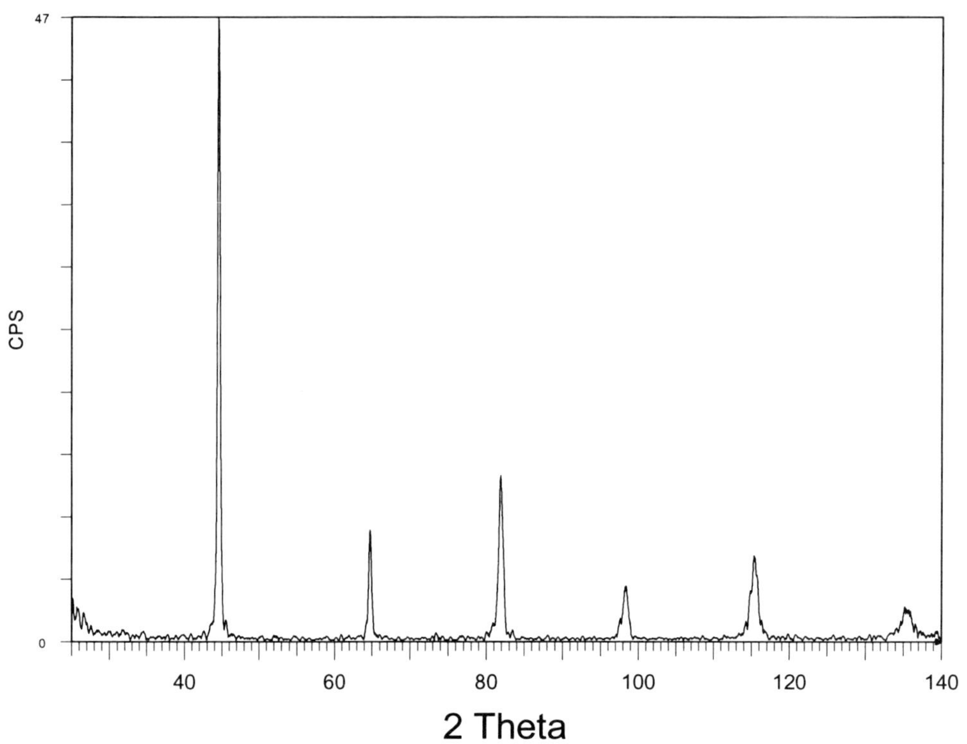

FIG. 1.2. X-ray diffraction pattern of chromium.

TABLE 1.5. Work Table for Chromium

Material: Chromium Radiation: Cu Kα $\lambda_{K\alpha 1} = 0.154056$ nm

Peak #	2θ (°)	$\sin^2\theta$	$\frac{\sin^2\theta}{\sin^2\theta_{min}}$	$\frac{\sin^2\theta}{\sin^2\theta_{min}} \times$ **?**	$h^2 + k^2 + l^2$	*hkl*	*a* (nm)
1	44.41						
2	64.59						
3	81.77						
4	98.26						
5	115.34						
6	135.47						

TABLE 1.6a. Work Table for Chromium—Analytical Method

Material: Chromium   Radiation: Cu Kα $\lambda_{K\alpha 1} = 0.154056$ nm

Peak #	2θ (°)	$\sin^2\theta$	$\frac{\sin^2\theta}{2}$	$\frac{\sin^2\theta}{3}$	$\frac{\sin^2\theta}{4}$	$\frac{\sin^2\theta}{5}$	$\frac{\sin^2\theta}{6}$	$\frac{\sin^2\theta}{8}$
1	44.41							
2	64.59							
3	81.77							
4	98.26							
5	115.34							
6	135.47							

TABLE 1.6b. Work Table for Chromium—Analytical Method

Material: Chromium   Radiation: Cu Kα   A =

Peak #	$\sin^2\theta$	$\frac{\sin^2\theta}{A}$	$h^2 + k^2 + l^2$	*hkl*
1				
2				
3				
4				
5				
6				

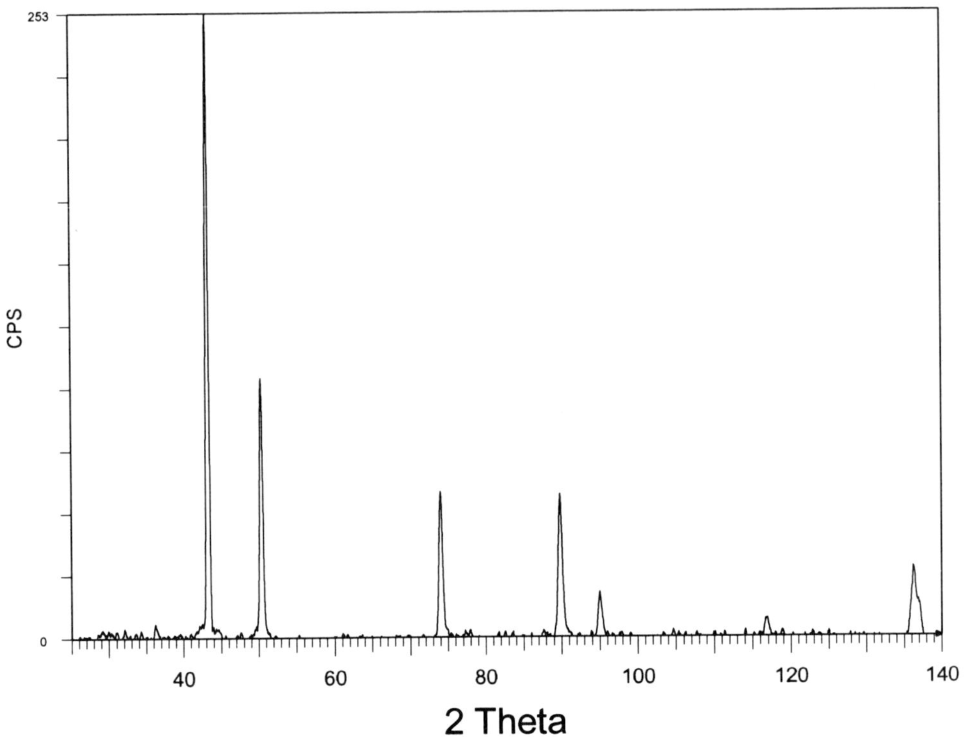

FIG. 1.3. X-ray diffraction pattern of copper.

TABLE 1.7. Work Table for Copper

Material: Copper | Radiation: Cu Kα | $\lambda_{K\alpha 1} = 0.154056$ nm

Peak #	2θ (°)	$\sin^2\theta$	$\frac{\sin^2\theta}{\sin^2\theta_{min}}$	$\frac{\sin^2\theta}{\sin^2\theta_{min}} \times$?	$h^2 + k^2 + l^2$	hkl	a (nm)
1	43.16						
2	50.30						
3	73.99						
4	89.85						
5	95.03						
6	116.92						
7	136.59						

TABLE 1.8a. Work Table for Copper—Analytical Method

Material: Copper Radiation: Cu Kα $\lambda_{K\alpha 1}$ = 0.154056 nm

Peak #	2θ (°)	$\sin^2\theta$	$\frac{\sin^2\theta}{2}$	$\frac{\sin^2\theta}{3}$	$\frac{\sin^2\theta}{4}$	$\frac{\sin^2\theta}{5}$	$\frac{\sin^2\theta}{6}$	$\frac{\sin^2\theta}{8}$
1	43.16							
2	50.30							
3	73.99							
4	89.85							
5	95.03							
6	116.92							
7	136.59							

TABLE 1.8b. Work Table for Copper—Analytical Method

Material: Copper Radiation: Cu Kα A =

Peak #	$\sin^2\theta$	$\frac{\sin^2\theta}{A}$	$h^2 + k^2 + l^2$	hkl
1				
2				
3				
4				
5				
6				
7				

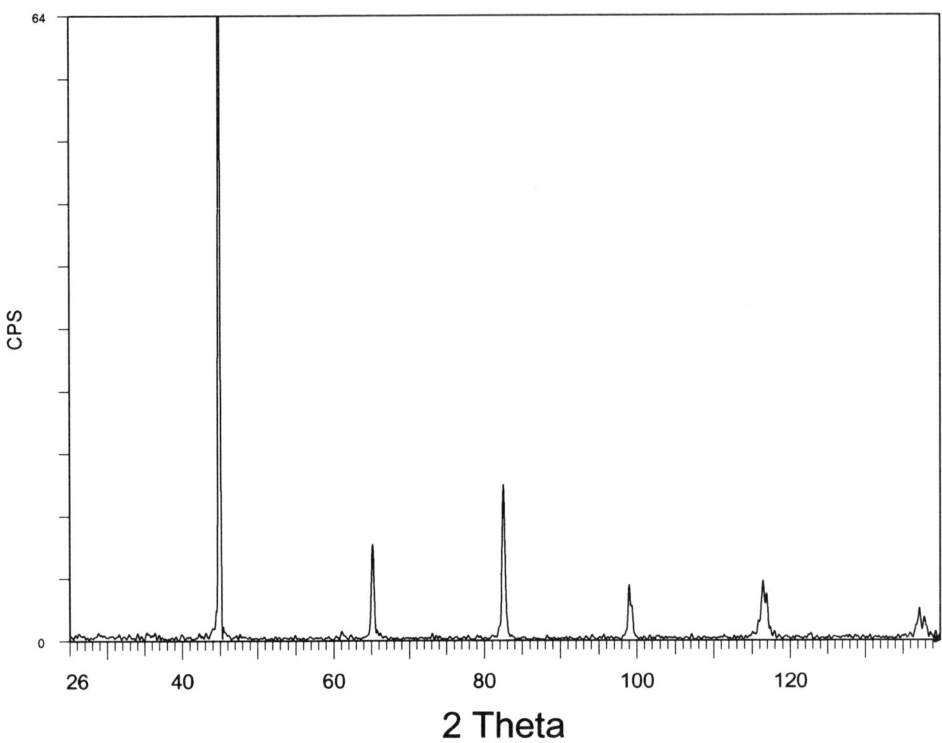

FIG. 1.4. X-ray diffraction pattern of iron.

TABLE 1.9. Work Table for Iron

Material: Iron Radiation: Cu Kα $\lambda_{K\alpha 1} = 0.154056$ nm

Peak #	2θ (°)	$\sin^2\theta$	$\dfrac{\sin^2\theta}{\sin^2\theta_{min}}$	$\dfrac{\sin^2\theta}{\sin^2\theta_{min}} \times ?$	$h^2+k^2+l^2$	*hkl*	*a* (nm)
1	44.73						
2	65.07						
3	82.40						
4	99.00						
5	116.45						
6	137.24						

TABLE 1.10a. Work Table for Iron—Analytical Method

Material: Iron Radiation: Cu Kα $\lambda_{K\alpha 1} = 0.154056$ nm

Peak #	2θ (°)	$\sin^2\theta$	$\frac{\sin^2\theta}{2}$	$\frac{\sin^2\theta}{3}$	$\frac{\sin^2\theta}{4}$	$\frac{\sin^2\theta}{5}$	$\frac{\sin^2\theta}{6}$	$\frac{\sin^2\theta}{8}$
1	44.73							
2	65.07							
3	82.40							
4	99.00							
5	116.45							
6	137.24							

TABLE 1.10b. Work Table for Iron—Analytical Method

Material: Iron Radiation: Cu Kα $A =$

Peak #	$\sin^2\theta$	$\frac{\sin^2\theta}{A}$	$h^2 + k^2 + l^2$	hkl
1				
2				
3				
4				
5				
6				

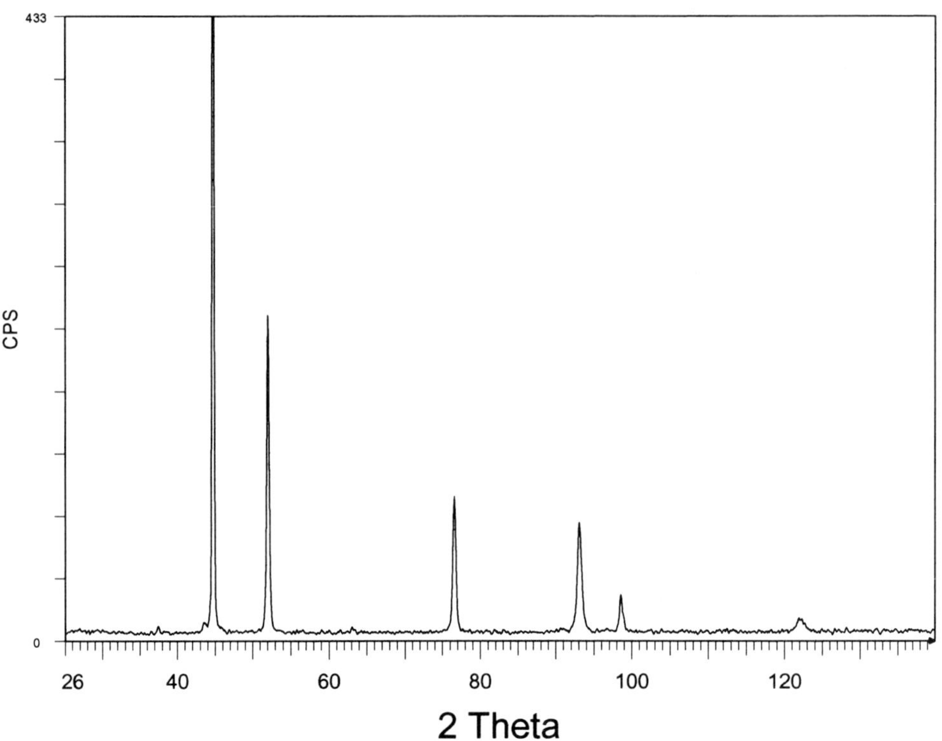

FIG. 1.5. X-ray diffraction pattern of nickel.

TABLE 1.11. Work Table for Nickel

Material: Nickel Radiation: Cu Kα $\lambda_{K\alpha 1} = 0.154056$ nm

Peak #	2θ (°)	$\sin^2\theta$	$\frac{\sin^2\theta}{\sin^2\theta_{min}}$	$\frac{\sin^2\theta}{\sin^2\theta_{min}} \times ?$	$h^2+k^2+l^2$	*hkl*	*a* (nm)
1	44.53						
2	51.89						
3	76.45						
4	93.01						
5	98.51						
6	122.12						

TABLE 1.12a. Work Table for Nickel—Analytical Method

Material: Nickel Radiation: Cu Kα $\lambda_{K\alpha 1} = 0.154056$ nm

Peak #	2θ (°)	$\sin^2\theta$	$\frac{\sin^2\theta}{2}$	$\frac{\sin^2\theta}{3}$	$\frac{\sin^2\theta}{4}$	$\frac{\sin^2\theta}{5}$	$\frac{\sin^2\theta}{6}$	$\frac{\sin^2\theta}{8}$
1	44.53							
2	51.89							
3	76.45							
4	93.01							
5	98.51							
6	122.12							

TABLE 1.12b. Work Table for Nickel—Analytical Method

Material: Nickel Radiation: Cu Kα $A =$

Peak #	$\sin^2\theta$	$\frac{\sin^2\theta}{A}$	$h^2 + k^2 + l^2$	hkl
1				
2				
3				
4				
5				
6				

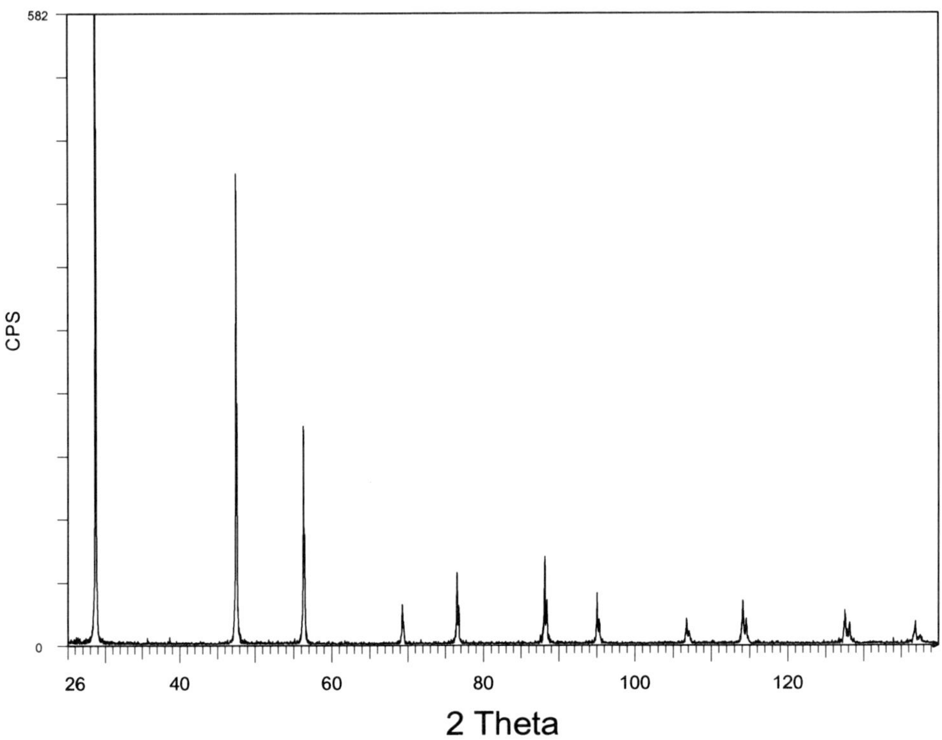

FIG. 1.6. X-ray diffraction pattern of silicon.

TABLE 1.13. Work Table for Silicon

Material: Silicon Radiation: Cu Kα $\lambda_{K\alpha 1} = 0.154056$ nm

Peak #	2θ (°)	$\sin^2\theta$	$\frac{\sin^2\theta}{\sin^2\theta_{min}}$	$\frac{\sin^2\theta}{\sin^2\theta_{min}} \times ?$	$h^2+k^2+l^2$	hkl	a (nm)
1	28.41						
2	47.35						
3	56.12						
4	69.10						
5	76.35						
6	88.04						
7	94.94						
8	106.75						
9	114.11						
10	127.56						
11	136.91						

TABLE 1.14a. Work Table for Silicon—Analytical Method

Material: Silicon Radiation: Cu Kα $\lambda_{K\alpha 1} = 0.154056$ nm

Peak #	2θ (°)	$\sin^2\theta$	$\frac{\sin^2\theta}{2}$	$\frac{\sin^2\theta}{3}$	$\frac{\sin^2\theta}{4}$	$\frac{\sin^2\theta}{5}$	$\frac{\sin^2\theta}{6}$	$\frac{\sin^2\theta}{8}$
1	28.41							
2	47.35							
3	56.12							
4	69.10							
5	76.35							
6	88.04							
7	94.94							
8	106.75							
9	114.11							
10	127.56							
11	136.91							

TABLE 1.14b. Work Table for Silicon—Analytical Method

Material: Silicon Radiation: Cu Kα $A =$

Peak #	$\sin^2\theta$	$\frac{\sin^2\theta}{A}$	$h^2 + k^2 + l^2$	*hkl*
1				
2				
3				
4				
5				
6				
7				
8				
9				
10				
11				

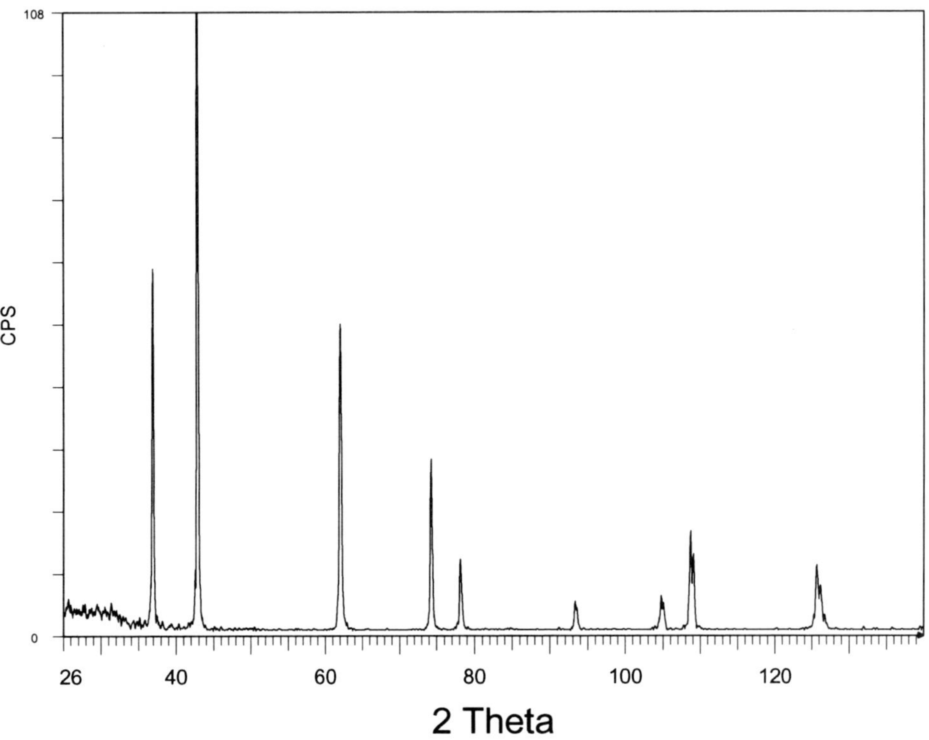

FIG. 1.7. X-ray diffraction pattern of titanium nitride.

TABLE 1.15. Work Table for Titanium Nitride

Material: Titanium nitride Radiation: Cu Kα $\lambda_{K\alpha 1} = 0.154056$ nm

Peak #	2θ (°)	$\sin^2\theta$	$\frac{\sin^2\theta}{\sin^2\theta_{min}}$	$\frac{\sin^2\theta}{\sin^2\theta_{min}} \times ?$	$h^2 + k^2 + l^2$	*hkl*	*a* (nm)
1	36.71						
2	42.67						
3	61.85						
4	74.12						
5	77.98						
6	93.30						
7	104.76						
8	108.61						
9	125.68						

TABLE 1.16a. Work Table for Titanium Nitride—Analytical Method

Material: Titanium nitride Radiation: Cu Kα $\lambda_{K\alpha 1} = 0.154056$ nm

Peak #	2θ (°)	$\sin^2 \theta$	$\frac{\sin^2 \theta}{2}$	$\frac{\sin^2 \theta}{3}$	$\frac{\sin^2 \theta}{4}$	$\frac{\sin^2 \theta}{5}$	$\frac{\sin^2 \theta}{6}$	$\frac{\sin^2 \theta}{8}$
1	36.71							
2	42.67							
3	61.85							
4	74.12							
5	77.98							
6	93.30							
7	104.76							
8	108.61							
9	125.68							

TABLE 1.16b. Work Table for Titanium Nitride—Analytical Method

Material: Chromium Radiation: Cu Kα A =

Peak #	$\sin^2\theta$	$\frac{\sin^2\theta}{A}$	$h^2 + k^2 + l^2$	hkl
1				
2				
3				
4				
5				
6				
7				
8				
9				

TABLE 1.17. Summary Table for Experimental Module 1

Material	Bravais lattice	Crystal structure	Lattice parameter (nm)
Aluminum	face-centered cubic	fcc	0.4049
Chromium			
Copper			
Iron			
Nickel			
Silicon			
Titanium nitride			

EXERCISES

1.1. Examine the diffraction patterns of copper and aluminum, and note that the 2θ values of reflections with the same Miller indices are shifted to lower 2θ values as you move from copper to aluminum. Explain why this occurs.

1.2. Schematically show how the diffraction peaks from a copper specimen (fcc, $a = 0.3615$ nm) shift in their angular positions if Mo Kα ($\lambda = 0.0711$ nm), Cu Kα ($\lambda = 0.1542$ nm), and Cr Kα ($\lambda = 0.2291$ nm) radiations are used to record the diffraction patterns.

1.3. A Te–Au alloy rapidly quenched from the liquid state is known to have a simple cubic structure with a lattice parameter of $a = 0.2961$ nm. Calculate the 2θ values for the different allowed reflections if the diffraction pattern were recorded with Cu Kα radiation. Represent these values in the form of a schematic diffraction pattern.

1.4. Both aluminum and titanium nitride have the face-centered cubic Bravais lattice, but with different lattice parameters. Even though the sequence of reflections is the same in both cases, the relative intensities of the reflections in the two patterns are quite different. Explain how the structures of these two materials differ and why the intensities of the reflections in the diffraction pattern are different.

1.5. Germanium has the diamond cubic structure with lattice parameter $a = 0.5658$ nm. Calculate the positions (2θ values) of the first six peaks expected in the x-ray diffraction pattern, assuming that Cu Kα radiation is used.

EXPERIMENTAL MODULE 2

Crystal Structure Determination. II: Hexagonal Structures

OBJECTIVE OF THE EXPERIMENT

To index the x-ray diffraction pattern and calculate the lattice parameters of some common materials with a hexagonal structure.

MATERIALS REQUIRED

Fine powders (–325 mesh, <45 μm in size) of magnesium (Mg), titanium (Ti), and aluminum nitride (AlN).

BACKGROUND AND THEORY

Many materials, including several metals, semiconductors, ceramics, and crystalline polymers, have a crystal structure based on the hexagonal Bravais lattice. The hexagonal unit cell is characterized by lattice parameters a and c. The plane spacing equation for the hexagonal structure is

$$\frac{1}{d^2} = \frac{4}{3}\left(\frac{h^2 + hk + k^2}{a^2}\right) + \frac{l^2}{c^2} \tag{2.1}$$

Combining Bragg's law ($\lambda = 2d \sin \theta$) with Eq. (2.1) yields

If the axial ratio of the material is not known, its value can be found by graphical methods, such as Hull–Davey charts. A description of these methods is beyond the scope of this book, and you are advised to look in the books by Klug and Alexander or Cullity listed in the Bibliography.

$$\frac{1}{d^2} = \frac{4}{3}\left(\frac{h^2 + hk + k^2}{a^2}\right) + \frac{l^2}{c^2} = \frac{4 \sin^2 \theta}{\lambda^2} \tag{2.2}$$

Rearranging gives

$$\sin^2 \theta = \frac{\lambda^2}{4}\left[\frac{4}{3}\left(\frac{h^2 + hk + k^2}{a^2}\right) + \frac{l^2}{c^2}\right] \tag{2.3}$$

Indexing the Pattern

In this experiment we assume that the c/a value (known as the axial ratio) of the hexagonal material is known (even though the patterns can be indexed without this knowledge).

We can rewrite Eq. (2.3) as

$$\sin^2 \theta = \frac{\lambda^2}{4a^2}\left[\frac{4}{3}\left(h^2 + hk + k^2\right) + \frac{l^2}{(c/a)^2}\right] \tag{2.4}$$

Equation (2.4) suggests that the $\sin^2 \theta$ value for each peak in the diffraction pattern is determined by the a and c lattice parameters of the unit cell and the Miller indices hkl. Since a and c/a are constants for a given diffraction pattern, according to Eq. (2.4) the $\sin^2 \theta$ values consist of two components, one involving the h and k values and the other involving the l value.

The values of $\frac{4}{3}(h^2 + hk + k^2)$ for different values of h and k are presented in Table 2.1. These values do not depend on the lattice parameters of the material.

TABLE 2.1. Values of $\frac{4}{3}(h^2 + hk + k^2)$

h	k	0	1	2	3
0		0	1.3333	5.3333	12.0000
1		1.3333	4.0000	9.3333	17.3333
2		5.3333	9.3333	16.0000	25.3333
3		12.0000	17.3333	25.3333	36.0000

TABLE 2.2. Values of $l^2/(c/a)^2$ for Zinc ($c/a = 1.8563$)

l	l^2	$l^2/(c/a)^2$
0	0	0.0000
1	1	0.2902
2	4	1.1608
3	9	2.6118
4	16	4.6432
5	25	7.2550
6	36	10.4472

Values of $l^2/(c/a)^2$ (to be calculated for each material with a different c/a value) can be tabulated as shown in Table 2.2. Let's consider the example of zinc ($c/a = 1.8563$) to illustrate the calculations.

Now we can add the value of $l^2/(c/a)^2$ from Table 2.2 to $\frac{4}{3}(h^2 + hk + k^2)$ from Table 2.1 and arrange the possible combinations (i.e., those permitted by the structure factor) of $\frac{4}{3}(h^2 + hk + k^2) + l^2/(c/a)^2$ values for different *hkl* indices in increasing order. The structure factor equation for a hexagonal structure (with more than one atom per unit cell) suggests that reflections are *absent* when $h + 2k = 3N$ (where N is an integer) *and* l is odd; all other reflections will be present (see Sec. 2.8 in Part I for the general structure factor equation). A reflection is absent only when *both* of these conditions are satisfied. Thus, you must consider whether a reflection will be present or absent only when l is odd. If l is odd, check whether $h + 2k = 3N$; if so, the reflection will be absent. For example, even though both 301 and 103 reflections have an odd value of l, the 301 reflection will be absent ($h + 2k = 3 + 2 \times 0 = 3$, i.e., a multiple of three) and while the 103 reflection will be present ($h + 2k = 1 + 2 \times 0 = 1 \neq 3$, i.e., not a multiple of three). All reflections will be present if l is even. Since the a and c parameters are different for hexagonal metals, the h and l values in the Miller indices cannot be interchanged (but h and k can be interchanged since $a_1 = a_2$, as shown in Fig. 22c in Part I). Realize that it is possible to interchange the h, k, and l values in materials with a cubic structure since the unit-cell parameters are the same in all three directions. This has an important bearing on the multiplicity factor and, hence, on the intensity of the reflections, which we discussed in Part I.

The listing of $\frac{4}{3}(h^2 + hk + k^2) + l^2/(c/a)^2$ for different values of *hkl* for zinc (we have remembered to delete the values for the forbidden reflections) is shown in Table 2.3. Since the $\frac{4}{3}(h^2 + hk + k^2) + l^2/(c/a)^2$ values in Table 2.3 were calculated for specific (hkl) planes, the *hkl* values for each

The hexagonal structure with one atom per unit cell (hP1) will be the primitive unit cell. This structure is not very common, although a few equilibrium phases and some metastable phases produced by rapid solidification from the melt or by the application of high pressure have been shown to have the "$HgSn_{6-10}$-type" hexagonal (one atom per unit cell) structure. For metals with a hexagonal structure and two atoms per unit cell, the structure is called hexagonal close-packed (hcp). However, with a greater number of atoms in the unit cell, as is common with many ceramics and semiconductors, the structure is not close-packed.

The structure factor equation for hexagonal materials with one atom per unit cell suggests that all possible reflections will be present. You may recall that a similar situation existed in the case of cubic structures. In fact, all possible reflections will be present in the diffraction pattern when the Bravais lattice is primitive and contains only one atom per unit cell, irrespective of the crystal system to which the material belongs.

of the reflections in an unknown specimen can be assigned by noting that the sequence of reflections will be the same as in Table 2.3 for a given value of c/a.

Calculation of the Lattice Parameters

Two approaches can be used to calculate the lattice parameters (a and c). First, we can calculate a by looking for reflections of the type $hk0$. When we substitute $l = 0$ in Eq. (2.4) we obtain

$$\sin^2\theta = \frac{\lambda^2}{4a^2}\left[\frac{4}{3}(h^2 + hk + k^2)\right] \tag{2.5}$$

or

$$a^2 = \frac{\lambda^2}{3\sin^2\theta}(h^2 + hk + k^2) \tag{2.6a}$$

or

$$a = \frac{\lambda}{\sqrt{3}\sin\theta}\sqrt{h^2 + hk + k^2} \tag{2.6b}$$

Once the value of a is known, c can be calculated since the value of c/a is known. Alternatively, we can look for the $00l$-type reflections and calculate c from Eq. (2.4). Since both h and k are equal to 0, Eq. (2.4) reduces to

TABLE 2.3. $\frac{4}{3}(h^2 + hk + k^2) + l^2/(c/a)^2$ for Different Values of hkl for Zinc

hkl	$\frac{4}{3}(h^2 + hk + k^2) + l^2/(c/a)^2$
002	1.1608
100	1.3333
101	1.6235
102	2.4941
103	3.9451
110	4.0000
004	4.6432
112	5.1608
200	5.3333
201	5.6235
104	5.9765
202	6.4941
203	7.9451
105	8.5883
114	8.6432
210	9.3333
211	9.6235
204	9.9765
006	10.4472
212	10.4941
106	11.7805
213	11.9451
300	12.0000
205	12.5883
302	13.1608

$$\sin^2\theta = \frac{\lambda^2}{4a^2}\left[\frac{l^2}{(c/a)^2}\right] = \frac{\lambda^2}{4c^2}l^2 \tag{2.7}$$

or

$$c^2 = \frac{\lambda^2}{4\sin^2\theta}l^2 \tag{2.8a}$$

or

$$c = \frac{\lambda}{2\sin\theta}l \tag{2.8b}$$

Once you have determined the a and c lattice parameters, calculate c/a and ensure that it is the same as the one given.

WORKED EXAMPLE

Let's now index the x-ray diffraction pattern of zinc recorded with Cu Kα radiation (Fig. 2.1). There are 20 peaks present in the diffraction pattern, and the 2θ values for these peaks are listed in Table 2.4. The $\sin^2 \theta$ values have been calculated, and the values of $\frac{4}{3}(h^2 + hk + k^2) + l^2/(c/a)^2$ listed in Table 2.3 are arranged in increasing order. Following the procedure described earlier, we have completely indexed the diffraction pattern as shown in Table 2.4 and determined the lattice parameters of zinc.

From these calculations it is clear that, for zinc, a = 0.2664 nm, c = 0.4947 nm, and c/a = 1.857, very close to the given value. From the a (or c) value, the c (or a) lattice parameter can be calculated from the hkl-type reflections. This c (or a) value can then be used to refine the a (or c) value, and by doing a few iterations like this the accuracy of both a and c values can be considerably improved. However, you need not do this iteration in the experimental modules unless a high accuracy is required in your calculations.

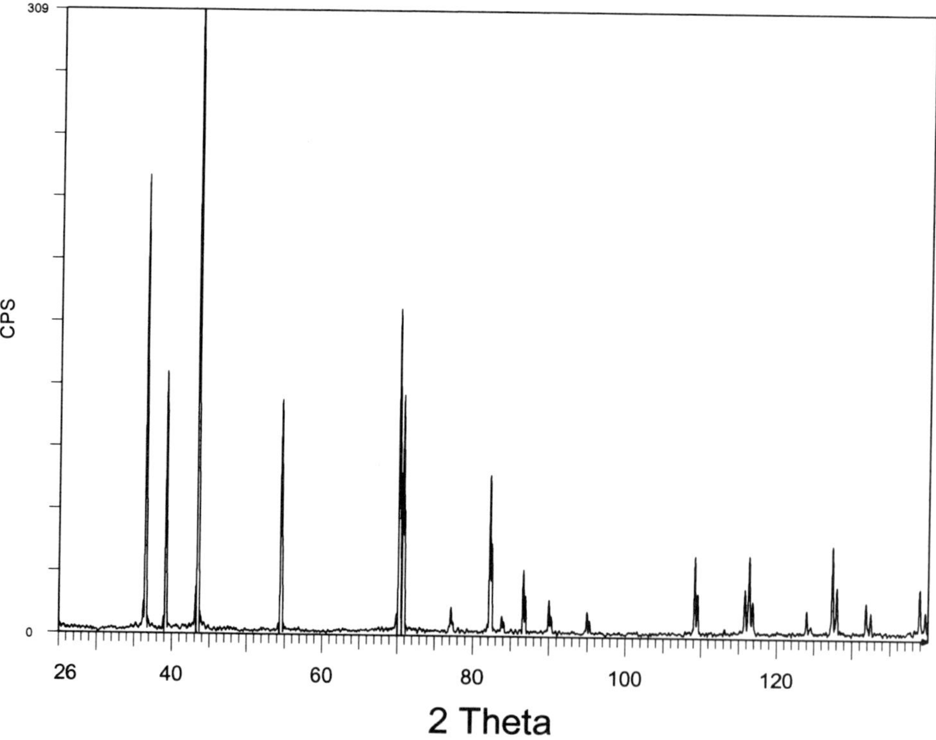

FIG. 2.1. X-ray diffraction pattern of zinc.

TABLE 2.4. Indexing the Diffraction Pattern of Zinc

Material: Zinc (c/a = 1.8563)			Radiation: Cu Kα	$\lambda_{K\alpha 1}$ = 0.154056 nm		
Peak #	2θ (°)	$\sin^2\theta$	$\frac{4}{3}(h^2+hk+k^2)+l^2/(c/a)^2$	*hkl*	*a* (nm)	*c* (nm)
1	36.31	0.0970	1.1608	002		0.4946
2	38.98	0.1113	1.3333	100	0.2665	
3	43.21	0.1356	1.6235	101		
4	54.32	0.2084	2.4941	102		
5	70.08	0.3296	3.9451	103		
6	70.64	0.3342	4.0000	110	0.2664	
7	77.04	0.3879	4.6432	004		0.4947
8	82.09	0.4312	5.1608	112		
9	83.72	0.4453	5.3333	200	0.2664	
10	86.54	0.4698	5.6235	201		
11	89.91	0.4992	5.9765	104		
12	94.88	0.5425	6.4941	202		
13	109.13	0.6639	7.9451	203		
14	115.80	0.7176	8.5883	105		
15	116.37	0.7221	8.6432	114		
16	124.03	0.7798	9.3333	210	0.2664	
17	127.47	0.8042	9.6235	211		
18	131.86	0.8337	9.9765	204		
19	138.56	0.8748	10.4472	006		0.4947
20	139.14	0.8781	10.4941	212		

ANALYTICAL METHOD

The foregoing procedure for indexing the diffraction pattern and determining the lattice parameters is time consuming and can sometimes be tricky in matching the *hkl* values to the observed $\sin^2\theta$ values. For example, if one of the peaks is very weak, it is possible that you may have missed or ignored that peak while measuring the 2θ values. This will offset all the *hkl* values and will give you the wrong values for the lattice parameters. Therefore, alternative analytical methods can be used to index diffraction patterns from materials with hexagonal structures (similar to the methods we described for cubic structures). These analytical methods will be now described.

We have that

$$\sin^2\theta = \frac{\lambda^2}{4a^2}\left[\frac{4}{3}(h^2+hk+k^2)+\frac{l^2}{(c/a)^2}\right] \tag{2.4}$$

Since a and c/a are constants for a given pattern, Eq. (2.4) can be rearranged as

$$\sin^2\theta = A(h^2 + hk + k^2) + Cl^2 \tag{2.9}$$

where $A = \lambda^2/3a^2$ and $C = \lambda^2/4c^2$. Since h, k, and l can only be integers, $h^2 + hk + k^2$ can have values like 0, 1, 3, 4, 7, 9, 12, . . . , as listed in Table 2.5, and l^2 can have values like 0, 1, 4, 9, ...

Now we divide the observed $\sin^2\theta$ values of the different peaks in the pattern by the integers 3, 4, 7, . . . and look for the common quotient, or that which is equal to one of the observed $\sin^2\theta$ values.

> You need to do this for the first few lines (say 10 to 15) in the pattern to get the trend and the common quotient. Once these are obtained the whole pattern can be subsequently indexed.

The $\sin^2\theta$ values representing this common value refer to the $hk0$-type reflections, and the quotient can be tentatively taken as A. Once the value of A is known, the value of C can be obtained by rearranging the terms in Eq. (2.9) as

$$Cl^2 = \sin^2\theta - A(h^2 + hk + k^2) \tag{2.10}$$

To obtain the value of C, we subtract from each $\sin^2\theta$ value the values of A, $3A$, $4A$, . . . and look for remainders that are in the ratio of 1, 4, 9, . . . (since l can only have integer values such as 1, 2, 3, . . . , l^2 will have values 1, 4, 9, . . .), which will be reflections of the type $00l$. From these, we can find the value of C. The remaining peaks that are neither $hk0$ type nor $00l$ type, belong to the hkl category. From a combination of A and C values, all the peaks in the diffraction pattern can be indexed. A final check on their correctness is made by comparing observed and calculated $\sin^2\theta$ values.

Once the values of A and C are known, the a and c lattice parameters of the unit cell can be calculated from the relationships $A = \lambda^2/3a^2$ and $C = \lambda^2/4c^2$.

TABLE 2.5. Values of $h^2 + hk + k^2$ for Different Values of h and k

hk[a]	00	10	11	20	21	30	22
$h^2 + hk + k^2$	0	1	3	4	7	9	12

[a] Remember that h and k are interchangeable and, therefore, 20 and 02 represent the same hk values.

WORKED EXAMPLE

Let's now index the x-ray diffraction pattern of zinc, using the analytical approach. In Table 2.6 we have divided the first 10 $\sin^2\theta$ values of the diffraction pattern by 3, 4, 7, and 9 to find the common quotient. It is clear that the common quotient, and hence the value of A is 0.1113. Once A is known, integral multiples of A ($3A$, $4A$, $7A$, . . .) can be identified and the $\sin^2\theta$ values corresponding to them will be from *hk*0-type reflections. For example, in such a pattern, $\sin^2\theta$ values of 0.3342 (= 3 × 0.1113) and 0.4453 (= 4 × 0.1113) correspond to 110 and 200 reflections, respectively. After this is done, reflections of type 00*l* can be identified by subtracting the values of A and $3A$ from the value of $\sin^2\theta$, as listed in Table 2.7.

You can see from Table 2.7 that 0.0971 is another common value. Since the 001 reflection is not allowed in the hcp structure, this value can be taken as the $\sin^2\theta$ value corresponding to the 002 reflection; i.e., $4C$ = 0.0971 or C = 0.0243. Also note that for peak 7, $\sin^2\theta = 0.3879$, which is nearly 16 times the C value, so this corresponds to the 004 reflection. Now, using the combination of A and C values we have obtained we indexed the whole pattern, as shown in Table 2.8. Note that sometimes you may be able to get $\sin^2\theta$ close to the observed value by more than one combination of the A and C values. In such a case you have to choose only that combination of *hkl* which is permitted by the structure factor.

The indexing shows that the observed and calculated $\sin^2\theta$ values match very well. Also note that the sequence of reflections is the same as

TABLE 2.6. Calculation of the Common Quotient A and Indexing the Pattern

Material: Zinc (c/a = 1.8563)			Radiation: Cu Kα			$\lambda_{K\alpha 1}$ = 0.154056 nm	
Peak #	2θ (°)	$\sin^2\theta$	$\frac{\sin^2\theta}{3}$	$\frac{\sin^2\theta}{4}$	$\frac{\sin^2\theta}{7}$	$\frac{\sin^2\theta}{9}$	*hkl*
1	36.31	0.0970	0.0323	0.0242	0.0139	0.0108	
2	38.98	**0.1113**	0.0371	0.0279	0.0159	0.0124	100
3	43.21	0.1356	0.0457	0.0343	0.0196	0.0152	
4	54.32	0.2084	0.0695	0.0521	0.0298	0.0231	
5	70.08	0.3296	0.1098	0.0824	0.0471	0.0366	
6	70.64	0.3342	**0.1114**	0.0846	0.0478	0.0371	110
7	77.04	0.3879	0.1293	0.0969	0.0555	0.0431	
8	82.09	0.4312	0.1437	0.1078	0.0616	0.0479	
9	83.72	0.4453	0.1486	**0.1113**	0.0637	0.0495	200
10	86.54	0.4698	0.1567	0.1175	0.0671	0.0522	

TABLE 2.7. Calculation of the C Value

Material: Zinc		Radiation: Cu Kα			$A = 0.1113$
Peak #	2θ (°)	$\sin^2\theta$	$\sin^2\theta - A$	$\sin^2\theta - 3A$	*hkl*
1	36.31	**0.0970**			002
2	38.98	0.1113	0.0000		
3	43.21	0.1356	0.0243		
4	54.32	0.2084	**0.0971**		102
5	70.08	0.3296	0.2183		
6	70.64	0.3342	0.2229		
7	77.04	0.3879	0.2766	0.0540	004
8	82.09	0.4312	0.3199	**0.0973**	112
9	83.72	0.4453	0.3340	0.1114	
10	86.54	0.4698	0.3585	0.1359	

TABLE 2.8. Indexing of the Diffraction Pattern of Zinc Using the Analytical Approach

Material: Zinc		Radiation: Cu Kα	$A = 0.1113$		$C = 0.0243$
Peak #	2θ (°)	$\sin^2\theta_{observed}$	$A + C$[a]	$\sin^2\theta_{calculated}$	*hkl*
1	36.31	0.0970	$0A + 4C$	0.0972	002
2	38.98	0.1113	$1A + 0C$	0.1114	100
3	43.21	0.1356	$1A + 1C$	0.1357	101
4	54.32	0.2084	$1A + 4C$	0.2086	102
5	70.08	0.3296	$1A + 9C$	0.3301	103
6	70.64	0.3342	$3A + 0C$	0.3342	110
7	77.04	0.3879	$0A + 16C$	0.3888	004
8	82.09	0.4312	$3A + 4C$	0.4314	112
9	83.72	0.4453	$4A + 0C$	0.4456	200
10	86.54	0.4698	$4A + 1C$	0.4699	201
11	89.91	0.4992	$1A + 16C$	0.5002	104
12	94.88	0.5425	$4A + 4C$	0.5428	202
13	109.13	0.6639	$4A + 9C$	0.6643	203
14	115.80	0.7176	$1A + 25C$	0.7189	105
15	116.37	0.7221	$3A + 16C$	0.7230	114
16	124.03	0.7798	$7A + 0C$	0.7798	210
17	127.47	0.8042	$7A + 1C$	0.8041	211
18	131.86	0.8337	$4A + 16C$	0.8344	204
19	138.56	0.8748	$0A + 36C$	0.8748	006
20	139.14	0.8781	$7A + 4C$	0.8770	212

[a]Combination of A and C

observed by the earlier method, suggesting that correct indexing can be done by either method.

Since $A = \lambda^2/3a^2$ and $C = \lambda^2/4c^2$, the a and c lattice parameters for zinc are $a = 0.2665$ nm and $c = 0.4941$ nm. These values are the same as those calculated earlier.

Note: While indexing the x-ray diffraction pattern of titanium, notice that the 210 reflection, permitted by the structure factor, is expected to be present. This reflection will have a low relative intensity. But in the standard pattern published by the JCPDS–International Centre for Diffraction Data (PDF #5-682), this reflection was *wrongly* omitted from the listing. The new corrected data are contained in PDF #44-1294 published in 1997. If your version of the software (or your copy of the PDF) was purchased prior to 1997 you have the old file.

EXPERIMENTAL PROCEDURE

Take fine powders (–325 mesh, <45 μm in size) of magnesium (Mg), titanium (Ti), and aluminum nitride (AlN). Load each powder separately

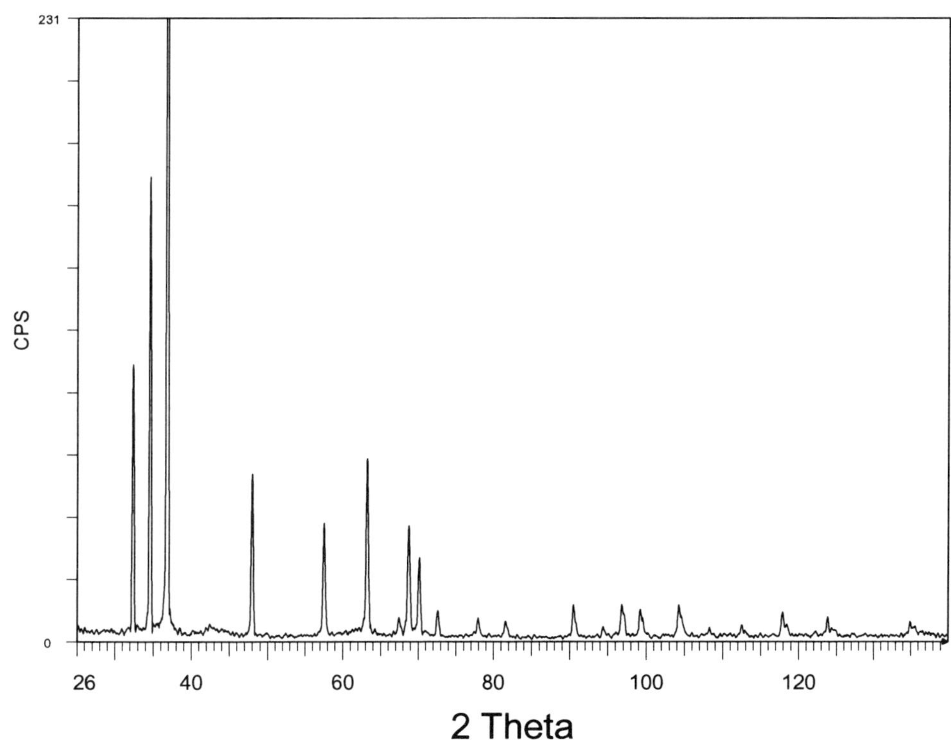

FIG. 2.2. X-ray diffraction pattern of magnesium.

TABLE 2.9. Values of c/a for the Different Materials Used

Material	Magnesium	Titanium	Aluminum Nitride
c/a	1.624	1.587	1.600

into the x-ray holder and record the x-ray diffraction patterns, using Cu Kα radiation in the 2θ angular range of 25 to 140°. Locate the maximum of each peak in the diffraction pattern and list the 2θ values in increasing order. Calculate the $\frac{4}{3}(h^2 + hk + k^2) + l^2/(c/a)^2$ values for each material from the c/a value given in Table 2.9 and arrange these values in increasing order. (*Remember*: Neglect the $\frac{4}{3}(h^2 + hk + k^2) + l^2/(c/a)^2$ values for the *hkl* reflections forbidden by the structure factor.) Index the diffraction pattern, using the two procedures described, and calculate the lattice parameters in each case. Even though it may be instructive to index the whole pattern, it is usually sufficient to index the major (most intense) reflections to evaluate the lattice parameters. We suggest that you index all the reflections. Tabulate your calculations in the work tables provided. (The x-ray diffraction patterns of Mg, Ti, and AlN are provided in Figs. 2.2, 2.3, and 2.4, respectively.)

RESULTS

Recall that the c/a value is different for different materials, so you need to calculate the $l^2/(c/a)^2$ values for each material whose diffraction pattern has to be indexed. You then add these values to $\frac{4}{3}(h^2 + hk + k^2)$ (which are

TABLE 2.10. Calculation of $l^2/(c/a)^2$ Values for Magnesium ($c/a = 1.624$)

l	l^2	$l^2/(c/a)^2$
0	0	0.0000
1	1	
2	4	
3	9	
4	16	
5	25	
6	36	

TABLE 2.11. Work Table for Magnesium

Material: Magnesium Radiation: Cu Kα $\lambda_{K\alpha 1} = 0.154056$ nm

Peak #	2θ (°)	$\sin^2 \theta$	$\frac{4}{3}(h^2 + hk + k^2) + l^2/(c/a)^2$	*hkl*	*a* (nm)	*c* (nm)
1	32.16					
2	34.38					
3	36.60					
4	47.80					
5	57.36					
6	63.05					
7	67.30					
8	68.61					
9	70.00					
10	72.46					
11	77.83					
12	81.51					
13	90.42					
14	94.31					
15	96.83					
16	99.21					
17	104.29					
18	104.56					
19	108.24					
20	112.49					

TABLE 2.12a. Work Table for Magnesium—Analytical Method

Material: Magnesium Radiation: Cu Kα $\lambda_{K\alpha 1} = 0.154056$ nm

Peak #	2θ (°)	$\sin^2\theta$	$\frac{\sin^2\theta}{3}$	$\frac{\sin^2\theta}{4}$	$\frac{\sin^2\theta}{7}$	$\frac{\sin^2\theta}{9}$	*hkl*
1	32.16						
2	34.38						
3	36.60						
4	47.80						
5	57.36						
6	63.05						
7	67.30						
8	68.61						
9	70.00						
10	72.46						
11	77.83						
12	81.51						
13	90.42						
14	94.31						
15	96.83						
16	99.21						
17	104.29						
18	104.56						
19	108.24						
20	112.49						

TABLE 2.12b. Work Table for Magnesium—Analytical Method

Material: Magnesium $\lambda_{K\alpha 1} = 0.154056$ nm $A =$

Peak #	2θ (°)	$\sin^2 \theta$	$\sin^2 \theta - A$	$\sin^2 \theta - 3A$	*hkl*
1	32.16				
2	34.38				
3	36.60				
4	47.80				
5	57.36				
6	63.05				
7	67.30				
8	68.61				
9	70.00				
10	72.46				
11	77.83				
12	81.51				
13	90.42				
14	94.31				
15	96.83				
16	99.21				
17	104.29				
18	104.56				
19	108.24				
20	112.49				

TABLE 2.12c. Work Table for Magnesium—Analytical Method

Material: Magnesium   Radiation: Cu Kα A =   C =

Peak #	2θ (°)	$\sin^2 \theta_{observed}$	$A + C$	$\sin^2 \theta_{calculated}$	hkl
1	32.16				
2	34.38				
3	36.60				
4	47.80				
5	57.36				
6	63.05				
7	67.30				
8	68.61				
9	70.00				
10	72.46				
11	77.83				
12	81.51				
13	90.42				
14	94.31				
15	96.83				
16	99.21				
17	104.29				
18	104.56				
19	108.24				
20	112.49				

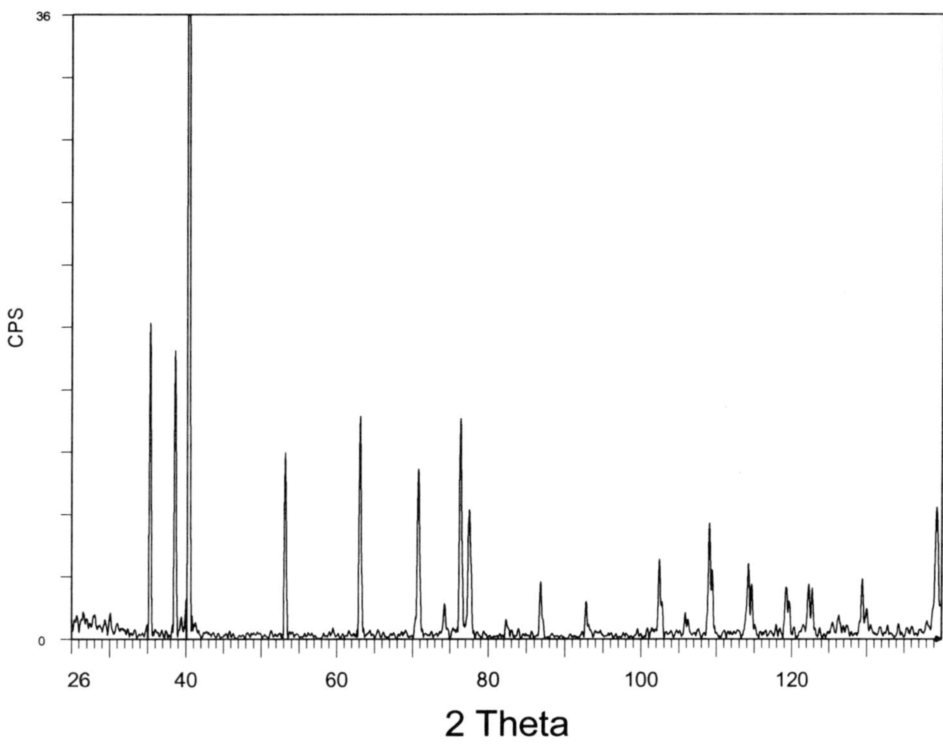

FIG. 2.3. X-ray diffraction pattern of titanium.

TABLE 2.13. Calculation of $l^2/(c/a)^2$ Values for Titanium ($c/a = 1.587$)

l	l^2	$l^2/(c/a)^2$
0	0	0.0000
1	1	
2	4	
3	9	
4	16	
5	25	
6	36	

TABLE 2.14. Work Table for Titanium

Material: Titanium Radiation: Cu Kα $\lambda_{K\alpha 1} = 0.154056$ nm

Peak #	2θ (°)	$\sin^2\theta$	$\frac{4}{3}(h^2 + hk + k^2) + l^2/(c/a)^2$	*hkl*	*a* (nm)	*c* (nm)
1	35.10					
2	38.39					
3	40.17					
4	53.00					
5	62.94					
6	70.65					
7	74.17					
8	76.21					
9	77.35					
10	82.20					
11	86.74					
12	92.68					
13	102.35					
14	105.60					
15	109.05					
16	114.22					
17	119.28					
18	122.30					
19	126.42					
20	129.48					

TABLE 2.15a. Work Table for Titanium—Analytical Method

Material: Titanium Radiation: Cu Kα $\lambda_{K\alpha 1} = 0.154056$ nm

Peak #	2θ (°)	$\sin^2\theta$	$\frac{\sin^2\theta}{3}$	$\frac{\sin^2\theta}{4}$	$\frac{\sin^2\theta}{7}$	$\frac{\sin^2\theta}{9}$	*hkl*
1	35.10						
2	38.39						
3	40.17						
4	53.00						
5	62.94						
6	70.65						
7	74.17						
8	76.21						
9	77.35						
10	82.20						
11	86.74						
12	92.68						
13	102.35						
14	105.60						
15	109.05						
16	114.22						
17	119.28						
18	122.30						
19	126.42						
20	129.48						

TABLE 2.15b. Work Table for Titanium—Analytical Method

Material: Titanium $\lambda_{K\alpha 1} = 0.154056$ nm $A =$

Peak #	2θ (°)	$\sin^2 \theta$	$\sin^2 \theta - A$	$\sin^2 \theta - 3A$	*hkl*
1	35.10				
2	38.39				
3	40.17				
4	53.00				
5	62.94				
6	70.65				
7	74.17				
8	76.21				
9	77.35				
10	82.20				
11	86.74				
12	92.68				
13	102.35				
14	105.60				
15	109.05				
16	114.22				
17	119.28				
18	122.30				
19	126.42				
20	129.48				

TABLE 2.15c. Work Table for Titanium—Analytical Method

Material: Titanium Radiation: Cu Kα $A =$ $C =$

Peak #	2θ (°)	$\sin^2 \theta_{observed}$	$A + C$	$\sin^2 \theta_{calculated}$	*hkl*
1	35.10				
2	38.39				
3	40.17				
4	53.00				
5	62.94				
6	70.65				
7	74.17				
8	76.21				
9	77.35				
10	82.20				
11	86.74				
12	92.68				
13	102.35				
14	105.60				
15	109.05				
16	114.22				
17	119.28				
18	122.30				
19	126.42				
20	129.48				

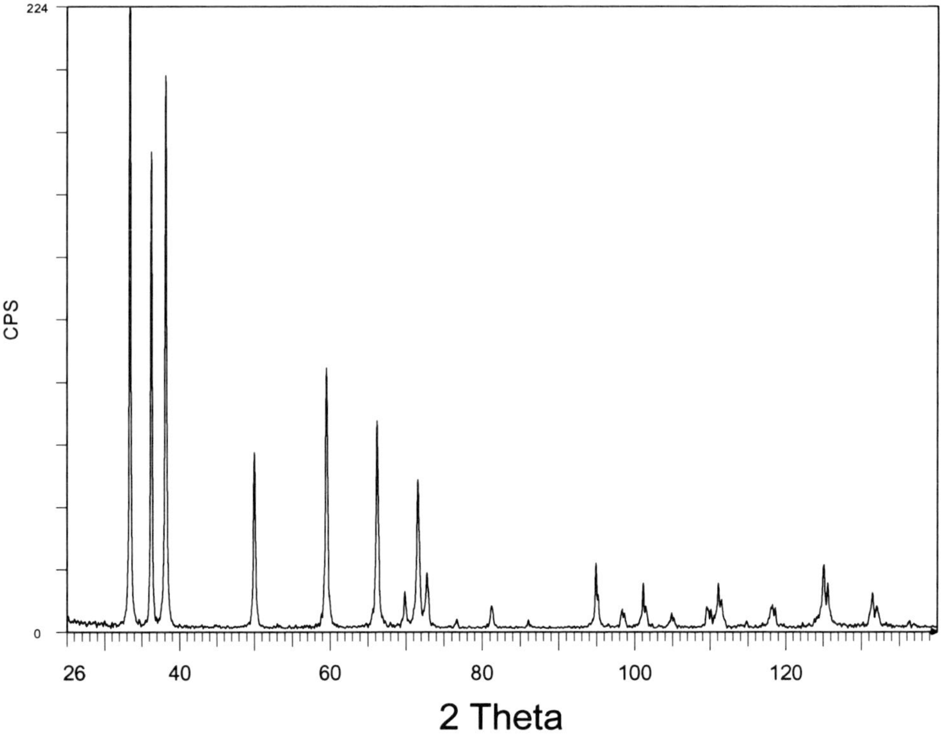

FIG. 2.4. X-ray diffraction pattern of aluminum nitride.

TABLE 2.16. Calculation of $l^2/(c/a)^2$ Values for Aluminum Nitride ($c/a = 1.600$)

l	l^2	$l^2/(c/a)^2$
0	0	0.0000
1	1	
2	4	
3	9	
4	16	
5	25	
6	36	

TABLE 2.17. Work Table for Aluminum Nitride

Material: Aluminum nitride Radiation: Cu Kα $\lambda_{K\alpha 1} = 0.154056$ nm

Peak #	2θ (°)	$\sin^2 \theta$	$\frac{4}{3}(h^2 + hk + k^2) + l^2/(c/a)^2$	*hkl*	*a* (nm)	*c* (nm)
1	33.20					
2	36.13					
3	38.00					
4	49.81					
5	59.39					
6	66.06					
7	69.68					
8	71.43					
9	72.60					
10	76.58					
11	81.14					
12	85.93					
13	94.82					
14	98.32					
15	101.01					
16	104.76					
17	109.55					
18	111.07					
19	114.81					
20	118.20					

TABLE 2.18a. Work Table for Aluminum Nitride—Analytical Method

Material: Aluminum nitride Radiation: Cu Kα $\lambda_{K\alpha 1} = 0.154056$ nm

Peak #	2θ (°)	$\sin^2\theta$	$\frac{\sin^2\theta}{3}$	$\frac{\sin^2\theta}{4}$	$\frac{\sin^2\theta}{7}$	$\frac{\sin^2\theta}{9}$	*hkl*
1	33.20						
2	36.13						
3	38.00						
4	49.81						
5	59.39						
6	66.06						
7	69.68						
8	71.43						
9	72.60						
10	76.58						
11	81.14						
12	85.93						
13	94.82						
14	98.32						
15	101.01						
16	104.76						
17	109.55						
18	111.07						
19	114.81						
20	118.20						

TABLE 2.18b. Work Table for Aluminum Nitride—Analytical Method

Material: Aluminum nitride $\lambda_{K\alpha 1} = 0.154056$ nm $A =$

Peak #	2θ (°)	$\sin^2 \theta$	$\sin^2 \theta - A$	$\sin^2 \theta - 3A$	*hkl*
1	33.20				
2	36.13				
3	38.00				
4	49.81				
5	59.39				
6	66.06				
7	69.68				
8	71.43				
9	72.60				
10	76.58				
11	81.14				
12	85.93				
13	94.82				
14	98.32				
15	101.01				
16	104.76				
17	109.55				
18	111.07				
19	114.81				
20	118.20				

TABLE 2.18c. Work Table for Aluminum Nitride—Analytical Method

Material: Aluminum nitride   Radiation: Cu Kα $A =$ $C =$

Peak #	2θ (°)	$\sin^2 \theta_{observed}$	$A + C$	$\sin^2 \theta_{calculated}$	*hkl*
1	33.20				
2	36.13				
3	38.00				
4	49.81				
5	59.39				
6	66.06				
7	69.68				
8	71.43				
9	72.60				
10	76.58				
11	81.14				
12	85.93				
13	94.82				
14	98.32				
15	101.01				
16	104.76				
17	109.55				
18	111.07				
19	114.81				
20	118.20				

TABLE 2.19. Summary Table for Experimental Module 2

Material	Lattice Parameters		
	a (nm)	c (nm)	c/a
Magnesium			
Titanium			
Aluminum nitride			

independent of the lattice parameters and hence are constant for all materials) to get the $\frac{4}{3}(h^2 + hk + k^2) + l^2/(c/a)^2$ values. Then arrange them in increasing order. Use the $\frac{4}{3}(h^2 + hk + k^2)$ values listed in Table 2.1 and calculate the $l^2/(c/a)^2$ values for each material and record them in Tables 2.10, 2.13, and 2.16. When you have finished the calculations, complete the summary Table 2.19.

EXERCISES

2.1. Compare indexing the x-ray diffraction patterns of magnesium ($c/a = 1.624$) and zinc ($c/a = 1.856$) and observe that the sequence of reflections is different in the two cases. Explain why this is so.

2.2. Zinc sulfide (ZnS) is known to exist in two modifications (or polymorphs). One form has the zinc blende structure (face-centered cubic Bravais lattice; a = 0.5406 nm), and the other form has the wurtzite structure (hexagonal Bravais lattice; a = 0.3821 nm, c = 0.6257 nm, and c/a = 1.6375). Calculate the expected diffraction patterns (2θ values) in the range of 25°–100° in both cases, assuming that Cu Kα radiation (λ = 0.1542 nm) is used. Explain why there are more reflections in the pattern for the hexagonal structure.

2.3. The fcc and hcp structures can be considered to be layered structures with a sequence of three layers (of close-packed planes) (ABCABCA . . .) for fcc and two layers (ABABABABA . . .) for hcp structures. These two arrangements give the same atomic packing factor when the axial ratio (c/a) for the hcp structure is ideal (c/a = 1.633), and it should be possible to transform one structure into the other. If the fcc structure has a lattice parameter of 0.400 nm, calculate the lattice parameters of the hcp structure and the diffraction patterns in both cases. Note that some reflections are common to both structures.

2.4. Polytetrafluoroethylene (PTFE, or Teflon as it is commonly known) is an example of a polymer that shows a crystalline conformation. The crystal structure of PTFE is hexagonal with $a = 0.566$ nm and $c = 1.950$ nm. Calculate the positions of the first six peaks expected in the x-ray diffraction pattern recorded from PTFE, using Cu Kα radiation.

EXPERIMENTAL MODULE 3

Precise Lattice Parameter Measurements

OBJECTIVE OF THE EXPERIMENT

To determine the precise lattice parameters of materials by x-ray diffraction methods.

MATERIALS REQUIRED

Fine silicon powder (–325 mesh, <45 μm in size).

BACKGROUND AND THEORY

A very precise knowledge of the lattice parameter(s) of materials is important in solid-state investigations. For example, such data have been of great help in developing satisfactory concepts of bonding energies in crystalline solids. Further, precise lattice parameter measurements are useful in determining solid solubility limits of one component in another (see Experimental Module 4), coefficients of thermal expansion, and true densities of materials. You should realize that in most of these cases, the magnitudes of changes in the lattice parameter(s) due to a change in solute content or temperature are so small that rather precise lattice parameter measurements must be made in order to measure these quantities with any accuracy. For example, the coefficients of thermal expansion (α) of most materials are about 10^{-6}/°C, and to determine small changes in lattice parameters as a function of temperature the lattice parameters must be measured to a very high accuracy. For aluminum α is $23.6 \times 10^{-6}/°C$, and its lattice parameter at 25°C is 0.4049 nm. If a piece of aluminum is heated to 50°C (an increase of 25°C), the lattice parameter increases by only 2.4×10^{-4} nm. To detect this change we require an accuracy of 0.06% in the measurement of the lattice parameter. On the other hand, if $\alpha < 23.6 \times 10^{-6}/°C$, as is often the case, especially for many ceramics and semiconductors, the accuracy needed in the measurement of lattice parameters is even higher.

Here we must distinguish between the terms "precision" and "accuracy." In everyday speech, we often use these terms interchangeably, but in scientific measurements they have distinct meanings. *Precision* is reproducibility and refers to how close the measurements in a series are to each other. *Accuracy*, on the other hand, refers to how close a measurement is to the "true" value. Thus, we may make highly precise but inaccurate measurements through the use of improper methods or equipment. High accuracy can be achieved only when the equipment is well calibrated, but good precision can be achieved with any equipment provided we take proper precautions in recording and analyzing the data. Therefore, in this experiment we assume that the equipment is in "ideal condition" and then consider methods to obtain the best precision in the lattice parameter(s) values.

Measurement of lattice parameters is an indirect process. For example, for a material with a cubic structure,

$$a = d\sqrt{h^2 + k^2 + l^2} \tag{3.1}$$

and d is measured from the sin θ value in the Bragg equation:

$$\lambda = 2d \sin\theta \tag{3.2}$$

Precision in measurement of a or d depends on precision in sin θ, which is a derived quantity, not on precision in θ, which is the measured quantity. (Remember that sin θ appears in the Bragg equation, not θ.) Figure 3.1 shows the variation of sin θ with θ, from which it may be noted that the error in measurement of sin θ decreases as the value of θ increases. For example, an error in θ of 1° leads to an error in sin θ of 1.7% at $\theta = 45°$ but only 0.15% at $\theta = 85°$. Therefore, the sin θ values will be reasonably accurate when θ is large even if θ itself is not measured very accurately.

The same result can be obtained by differentiating the Bragg equation with respect to θ, when we obtain

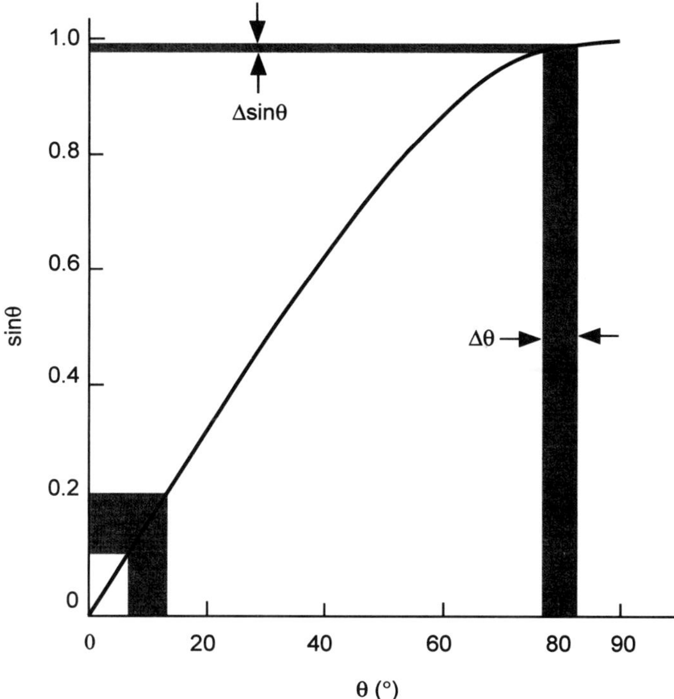

FIG. 3.1. Variation of sin θ with θ. The error in sin θ for a given error in θ decreases as the value of θ increases.

$$\frac{\Delta d}{d} = -\cot\theta\,\Delta\theta \tag{3.3}$$

Since cot θ = 0 when θ = 90°, the fractional error in *d* (i.e., $\Delta d/d$) is zero when θ = 90°. However, since we cannot observe an x-ray reflection at θ = 90° (this will be the back-reflected beam), we should consider reflections as close to the value of θ = 90° as possible to get the best precision and calculate the true value of the lattice parameter(s) as θ approaches 90°. Unfortunately, the plot of lattice parameter versus θ is not linear; therefore, extrapolation of the lattice parameter versus θ curve to θ = 90° does not lead to accurate results. However, extrapolation of the lattice parameter(s) against certain functions of θ will produce a straight line, which can then be extrapolated to the value corresponding to θ = 90°. Also note that the exact nature of this extrapolation function depends on the kind of equipment used to record the x-ray diffraction pattern (type of camera or diffractometer), as we will show.

Cameras are not widely used nowadays for x-ray diffraction studies since they involve photographic films and processing, which is time consuming; they also require the storage and disposal of chemicals. Additionally, film storage can be a problem when many analyses have been performed. We have included for completeness the method used for precise lattice parameter determination with a Debye–Scherrer camera, and at the end of this module you will find an exercise using this method. However, it is unlikely that you will encounter this method in practice.

Debye–Scherrer Cameras

Various effects that can lead to errors in the measurement of θ in the Debye–Scherrer camera are

- Film shrinkage
- Incorrect camera radius
- Off-centering of specimen
- Absorption in specimen
- Divergence of the beam

Some textbooks on x-ray diffraction (see the Bibliography) have discussed in detail the origin of these effects and how they can be corrected to achieve high precision in lattice parameter measurements. Considerations of the possible errors led to the development of extrapolation functions such as

$$\cos^2\theta \qquad \text{Bradley–Jay function}$$

$$\frac{\cos^2\theta}{\sin\theta} + \frac{\cos^2\theta}{\theta} \qquad \text{Nelson–Riley function}$$

When the lattice parameter is plotted against any of these extrapolation functions, a straight line can be drawn through the points. The lattice parameter corresponding to the value of 0 for these extrapolation functions (i.e., $\theta = 90°$) gives the true value of the lattice parameter, a_0, as shown in Fig. 3.2.

The Nelson–Riley extrapolation function has the advantage that it has a greater range of linearity and hence is useful when only a few high-angle reflections are present; reflections down to $\theta = 60°$ can also be used in this case. Further, the precision obtainable with the Nelson–Riley extrapolation function is better than what can be obtained with the Bradley–Jay extrapolation function.

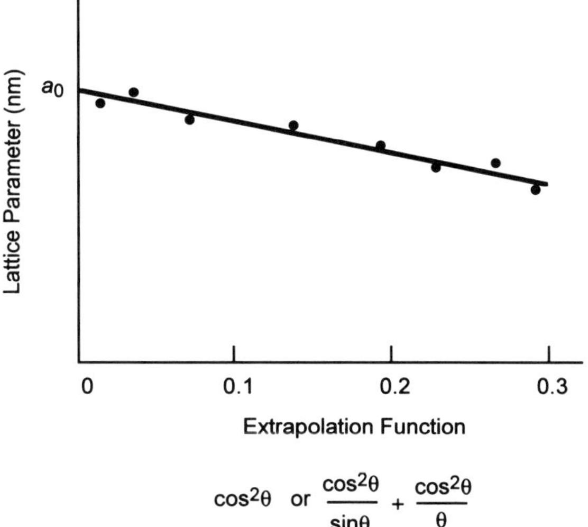

$$\cos^2\theta \quad \text{or} \quad \frac{\cos^2\theta}{\sin\theta} + \frac{\cos^2\theta}{\theta}$$

FIG. 3.2. The true value of the lattice parameter (a_0) is obtained by extrapolating the straight line of the lattice parameter versus an extrapolation function of θ to the value corresponding to 0, i.e., for θ = 90°.

Diffractometers

You are more likely to use a diffractometer than a camera to obtain an x-ray diffraction pattern. Use of diffractometers avoids many of the problems associated with cameras. Another big advantage is that the data can be stored on floppy disks and easily retrieved. It is also possible to produce high-quality versions (suitable for publication purposes!) of the diffraction patterns using, for example, laser printers. (All the x-ray diffraction patterns in this book have been produced in this way.) In addition, the data can also be directly processed by using computer software to obtain information about the structure, lattice parameters, lattice strain, crystallite size, etc., of the specimen.

A diffractometer is a more complex apparatus than a camera and therefore more subject to misalignment of its component parts. In a diffractometer, the important sources of error in calculating sin θ are

1. Misalignment of the instrument.
2. Use of a flat specimen instead of a specimen curved to conform to the focusing circle. This error is minimized, with loss of intensity, by decreasing the irradiated width of the specimen by means of an incident beam of small horizontal divergence.
3. Absorption in the specimen. The specimen thickness should be so chosen as to get reflections with the maximum intensity possible.

4. Displacement of the specimen from the diffractometer axis. This is usually the largest single source of error and causes an error in d given by

$$\frac{\Delta d}{d} = -\frac{D \cos^2 \theta}{R \sin \theta} \tag{3.4}$$

where D is the displacement of the specimen surface from the center of the diffractometer circle and R is the radius of the diffractometer circle.

5. Vertical divergence of the incident beam. This error is minimized, with loss of intensity, by decreasing the vertical opening of the receiving slit.

Since $\Delta d/d$ varies differently with different errors, a single extrapolation function is not satisfactory in this case. For example, $\Delta d/d$ varies as $\cos^2 \theta$ for errors 2 and 3 but as $\cos^2 \theta/\sin \theta$ for error 4. Therefore, the best way of deciding which of these errors is more significant would be to extrapolate the lattice parameter against $\cos^2 \theta/\sin \theta$ and against $\cos^2 \theta$ and see which function gives a better straight line and then decide which error is more significant.

No matter what method of recording the x-ray diffraction pattern you use, it is necessary to have as many reflections as possible in the high-angle region of the diffraction pattern so that you will have many points enabling you to draw the best possible straight line. When a peak is resolved into α_1 and α_2 components, you will have two lattice parameter points for each *hkl* value, one for each component of the resolved peak. The resolution of the peak into α_1 and α_2 components can be achieved by simply enlarging the 2θ scale. Further, the number of diffraction peaks in the pattern from a given material can be increased by

- Decreasing the wavelength of the radiation used (use Mo $K\alpha$ instead of Cu $K\alpha$), but this may not always be possible because most laboratories will have only one target.
- Using both the $K\alpha_1$ and $K\alpha_2$ components; this is easy to do because a well-annealed and reasonably large grained material will give a sharp and well-resolved diffraction pattern, and the angular separation between α_1 and α_2 increases with increasing value of θ.
- Using the $K\alpha$ ($K\alpha_1$ and $K\alpha_2$ components when they are resolved) and $K\beta$ components (i.e., unfiltered radiation); this also is not usually possible because modern diffractometers use monochromators that are aligned to diffract only the $K\alpha$ component.

Errors are of two kinds: random and systematic. Random errors are chance errors, such as those involved in measuring the position of the diffraction peak. These errors may be positive or negative, and they do not vary in a regular manner with some particular parameter, say the Bragg angle θ. Systematic errors, on the other hand, vary in a regular manner with θ, and these errors always have the same sign. For example, the value of the lattice parameter a always decreases as θ increases due to the systematic errors.

If Kα and Kβ components are both in the incident beam, then realize that all the crystal planes diffract the Kα and Kβ radiations. The presence of Kβ peaks in a pattern can usually be revealed by calculation, since if a certain set of planes reflect Kβ radiation at an angle θ_β, they must also reflect Kα radiation at an angle θ_α (unless θ_α exceeds 90°), and one angle may be calculated from the other from the relation

$$\frac{\lambda_{K\alpha}^2}{\lambda_{K\beta}^2}(\sin^2\theta_\beta) = \sin^2\theta_\alpha \tag{3.5}$$

where $\lambda_{K\alpha}^2/\lambda_{K\beta}^2$ has the value 1.226 for Cu K radiation and is near 1.2 for most radiations.

The best possible straight line through the several points can be drawn by using the least-squares method, which states that the most probable value of the measured quantity is that which makes the sum of the squares of the errors a minimum. Note that the random errors involved in measuring the peak positions are responsible for the deviation of the various points from the extrapolation line.

An analytical method that minimizes the random errors in a reproducible and objective manner has been proposed by Cohen (M. U. Cohen, *Rev. Sci. Instrum.* **6** (1935), 68; **7** (1936), 155), and it can be used to calculate the lattice parameters precisely for cubic and noncubic systems. The details of the calculation for the cubic system follow. (You can modify the calculations for noncubic systems by using the appropriate plane spacing equations listed in Appendix 1.)

Squaring the Bragg equation and taking logarithms, we get

$$\log\sin^2\theta = \log\left(\frac{\lambda^2}{4}\right) - 2\log d \tag{3.6}$$

Differentiating Eq. (3.6) gives

$$\frac{\Delta\sin^2\theta}{\sin^2\theta} = -\frac{2\,\Delta d}{d} \tag{3.7}$$

The straight line $y = a + bx$ should be fit through the experimental points $(x_1, y_1), (x_2, y_2), \ldots, (x_n, y_n)$ so that the sum of the squares of the distances of the points from the drawn straight line is a minimum, where the distance is measured in the y direction.

If we assume that the combined systematic errors take the form

$$\frac{\Delta d}{d} = K \cos^2 \theta \tag{3.8}$$

then combining Eqs. (3.7) and (3.8) gives

$$\Delta \sin^2 \theta = -2K \cos^2 \theta \sin^2 \theta = D \sin^2 2\theta \tag{3.9}$$

where D is a new constant. (This equation is valid only when the $\cos^2 \theta$ extrapolation function is valid. If some other extrapolation function is used, Eq. (3.8) must be modified accordingly.) The meaning of Eq. (3.9) is that the observed $\sin^2 \theta$ value for any given peak will be in error by the amount $D \sin^2 2\theta$ as a result of the combined action of the systematic errors.

The true value of $\sin^2 \theta$ for any diffraction peak is

$$\sin^2 \theta \text{ (true)} = \frac{\lambda^2}{4a_0^2}(h^2 + k^2 + l^2) \tag{3.10}$$

where a_0 is the true value of the lattice parameter we wish to obtain. But

$$\sin^2 \theta \text{ (observed)} - \sin^2 \theta \text{ (true)} = \Delta \sin^2 \theta \tag{3.11}$$

$$\sin^2 \theta \text{ (observed)} - \frac{\lambda^2}{4a_0^2}(h^2 + k^2 + l^2) = D \sin^2 2\theta \tag{3.12}$$

or

$$\sin^2 \theta \text{ (observed)} = A\alpha + C\delta \tag{3.13}$$

where $A = \lambda^2/4a_0^2$, $\alpha = (h^2 + k^2 + l^2)$, $C = D/10$, and $\delta = 10 \sin^2 2\theta$. (The factor 10 is introduced into the definition of C and δ solely to make the coefficients of the various terms in the normal equations of the same order of magnitude.) The parameter D is called the "drift" constant, and is a fixed quantity for any diffraction pattern but differs from one pattern to another. Best precision is achieved when the value of D is as small as possible. Equation (3.13) can be written for each reflection in the x-ray diffraction pattern.

In Eq. (3.13), we know $\sin^2\theta$ (observed), α from the indexing of the pattern, and δ. (Remember to use the observed θ values to calculate the δ values.) By solving the simultaneous equations for the observed reflections, we can calculate A and C. From A, the true value of the lattice parameter, a_0, can be calculated.

The foregoing procedure can be combined with the least-squares principle to minimize the effect of random observational errors. Because of these errors, Eq. (3.13) does not hold exactly for any particular reflection, but instead the quantity $A\alpha + C\delta - \sin^2\theta$ (observed) differs from zero by a small quantity, e. According to the theory of least squares, the best values of the coefficients A and C are those for which the sum of the squares of the random observational errors is a minimum; i.e.,

$$\sum (e)^2 = \text{a minimum} = \sum [A\alpha + C\delta - \sin^2\theta\ (\text{observed})]^2 \tag{3.14}$$

A pair of normal equations can be obtained by differentiating Eq. (3.14) with respect to A and C and equating them to zero. Thus,

$$\sum \alpha \sin^2\theta = A \sum \alpha^2 + C \sum \alpha\,\delta \tag{3.15}$$

$$\sum \delta \sin^2\theta = A \sum \alpha\delta + C \sum \delta^2 \tag{3.16}$$

By solving these equations we determine A, and from this value of A the true lattice parameter a_0 can be determined.

WORKED EXAMPLE

Let's now determine the lattice parameter of aluminum, an fcc metal, using both the extrapolation function and Cohen's least-squares methods. The x-ray diffraction pattern in Fig. 3.3 was recorded at 25°C with Cu Kα radiation. Since we are interested in the precise lattice parameter, we need to concentrate only on high-angle peaks. Accordingly, we consider the three high-angle peaks (in the 2θ angular range of 110 to 150°), which have been resolved into α_1 and α_2 components. The data are summarized in Table 3.1.

We can now plot the lattice parameters calculated against the $(\cos^2\theta)/(\sin\theta)$ extrapolation function, as shown in Fig. 3.4. By drawing the best possible straight line through these points, we find that the true lattice parameter of aluminum is $a_0 = 0.404929$ nm.

To use Cohen's analytical method, we need values of α, δ, α^2, $\alpha\delta$, and δ^2, and we need to sum the α^2, $\alpha\delta$, and δ^2, $\alpha \sin^2\theta$, and $\delta \sin^2\theta$ values to

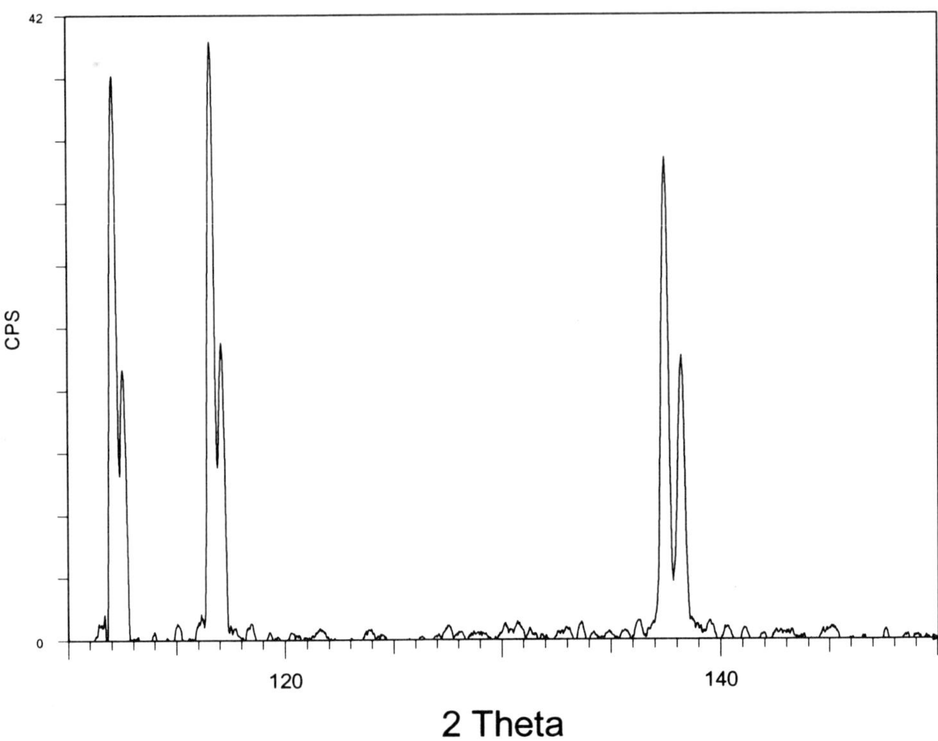

FIG. 3.3. X-ray diffraction pattern of aluminum showing only the high-angle region.

TABLE 3.1. Calculation of the Lattice Parameter of Aluminum

Material: Aluminum					Radiation: Cu Kα	
Peak #	θ (°)	λ	$\sin^2\theta$	$\sin^2\theta$ adjusted to $K\alpha_1$[a]	*hkl*	*a* (nm)
1	56.017	$K\alpha_1$	0.68758	0.68758	331	0.404915
2	56.232	$K\alpha_2$	0.69105	0.68763	331	0.404900
3	58.291	$K\alpha_1$	0.72374	0.72374	420	0.404922
4	58.523	$K\alpha_2$	0.72735	0.72375	420	0.404920
5	68.735	$K\alpha_1$	0.86846	0.86846	422	0.404929
6	69.107	$K\alpha_2$	0.87282	0.86849	422	0.404922

[a] $\sin^2\theta_{K\alpha1\,(adj)} = \sin^2\theta_{K\alpha2}\,(\lambda^2_{K\alpha1}/\lambda^2_{K\alpha2})$.

TABLE 3.2. Data Required for Cohen's Analytical Method

Material: Aluminum						Radiation: Cu Kα	
Peak #	α	δ	α^2	$\alpha\delta$	δ^2	$\alpha \sin^2 \theta$[a]	$\delta \sin^2 \theta$[a]
1	19	8.6	361	163.4	73.96	13.06402	5.91319
2	19	8.5	361	161.5	72.25	13.06497	5.84486
3	20	8.0	400	160.0	64.00	14.47700	5.79080
4	20	7.9	400	158.0	62.41	14.47500	5.71763
5	24	4.6	576	110.4	21.16	20.84280	3.99487
6	24	4.4	576	105.6	19.36	20.84760	3.82206
			$\Sigma \alpha^2 = 2674$	$\Sigma \alpha\delta = 858.9$	$\Sigma \delta^2 =$ 313.14	$\Sigma \alpha \sin^2 \theta =$ 96.77139	$\Sigma \delta \sin^2 \theta =$ 31.08341

[a]Remember to use the $\sin^2 \theta$ values adjusted to the Kα_1 wavelength.

formulate the normal equations. These values are listed in Table 3.2. The normal equations are

$$A \sum \alpha^2 + C \sum \alpha\delta = \sum \alpha \sin^2 \theta \tag{3.15}$$

$$A \sum \alpha\delta + C \sum \delta^2 = \sum \delta \sin^2 \theta \tag{3.16}$$

Substituting the appropriate values from Table 3.2, we get

$$2674A + 858.9C = 96.77139$$

$$858.9A + 313.14C = 31.08341$$

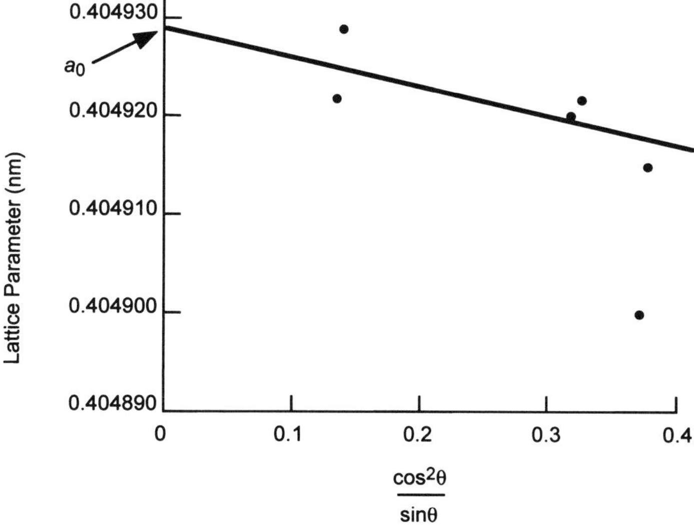

FIG. 3.4. Extrapolation of the calculated lattice parameter of aluminum against $(\cos^2 \theta)/(\sin \theta)$.

Solving the equations gives $A = 0.0361895$ and $C = 7.8 \times 10^{-7}$. Hence, the true lattice parameter for aluminum is $a_0 = \lambda/2\sqrt{A} = 0.40491$ nm.

Also note that $C = 7.8 \times 10^{-7}$ and $D = 10C = 7.8 \times 10^{-6}$, a very small value. Hence, we have a highly precise value of the lattice parameter.

EXPERIMENTAL PROCEDURE

Take fine silicon powder and record the x-ray diffraction pattern with Cu Kα radiation in the 2θ angular range of 100 to 160°. (We have provided the pattern in Fig. 3.5 if you are not able to record your own.) Make sure that the α_1 and α_2 components are well resolved in the diffraction pattern. Index the reflections in the diffraction pattern, using the procedure described in Experimental Module 1. (Remember that if you have already indexed the diffraction pattern of silicon in Experimental Module 1, you need not start from the beginning!) Calculate the lattice parameter for each reflection (for the α_1 and α_2 components) and plot the lattice

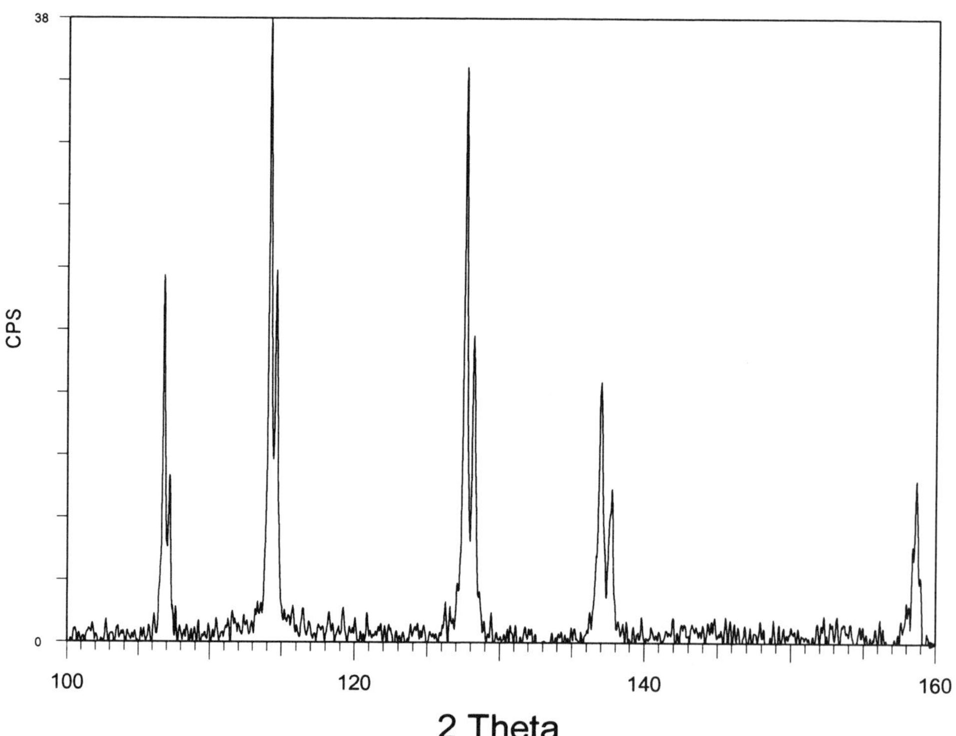

FIG. 3.5. X-ray diffraction pattern of silicon showing only the high-angle region.

While calculating the lattice parameter from different reflections, you should use the wavelengths of the α_1 and α_2 components listed in Table 2, Part I. Once again, for Cu Kα these values are $\lambda_{K\alpha 1} = 0.154056$ nm and $\lambda_{K\alpha 2} = 0.154439$ nm.

parameter against $\cos^2\theta/\sin\theta$ and against $\cos^2\theta$. Find the lattice parameter corresponding to the value of 0 for the given extrapolation functions (i.e., at $\theta = 90°$). Compare your value with that listed in standard reference books ($a_0 = 0.5431$ nm). How close were you?

RESULTS

Show your calculations in Table 3.3. The lattice parameter of silicon calculated from the extrapolation functions is

$\cos^2\theta$ function nm

TABLE 3.3. Work Table for Indexing the Diffraction Pattern of Silicon

Material: Silicon Radiation: Cu Kα

Peak #	2θ (°)	λ	*hkl*	a (nm)	$\cos^2\theta$	$(\cos^2\theta)/(\sin\theta)$
1	106.723	Kα_1				
2	107.127	Kα_2				
3	114.101	Kα_1				
4	114.551	Kα_2				
5	127.556	Kα_1				
6	128.141	Kα_2				
7	136.933	Kα_1				
8	137.669	Kα_2				
9	158.684	Kα_1				

$\frac{\cos^2\theta}{\sin\theta}$ function nm

EXERCISES

3.1. Calculate the lattice parameter of silicon (from the observed 2θ values in your experiment), using Cohen's least-squares method. How does this value compare with that determined by the extrapolation methods?

3.2. The following data were obtained from a Debye–Scherrer pattern of lead, an fcc metal, made with Cu Kα radiation.

$h^2 + k^2 + l^2$	35	35	36	36	40	40
λ	Kα_1	Kα_2	Kα_1	Kα_2	Kα_1	Kα_2
θ	67.080	67.421	69.061	69.467	79.794	80.601

Determine the lattice parameter, accurate to four significant figures, by graphical extrapolation of the lattice parameter a against $\cos^2\theta$ and against $(\cos^2\theta)/(\sin\theta) + (\cos^2\theta)/\theta$. Also calculate the lattice parameter by using Cohen's least-squares method.

EXPERIMENTAL MODULE 4

Phase Diagram Determination

OBJECTIVE OF THE EXPERIMENT

To determine the solid-state portion of a binary phase diagram by x-ray diffraction methods.

MATERIALS REQUIRED

Fine powders (–325 mesh, <45 μm in size) of MgO and NiO and reacted mixtures of MgO with 20, 40, 60, and 80 mol% NiO.

BACKGROUND AND THEORY

Phase diagrams are very useful in materials science because they establish the phase fields of a system (i.e., which phases are present) at different temperatures and compositions (and pressures). Many phase diagrams show the presence of primary (or terminal) solid solutions, and some also show the occurrence of secondary solid solutions (or intermediate phases). Three typical phase diagrams are shown in Fig. 4.1. Figure 4.1a shows an isomorphous system in which the two components, Ge and Si (both semiconductors), are completely soluble in each other in the liquid and solid states. Figure 4.1b shows a simple eutectic system in which the two components, MgO and CaO (both ceramics), have limited solubilities in the solid state but are completely soluble in the liquid state. A reacted mixture of MgO–CaO beyond the solid solubility level contains both the MgO and CaO solid solution (ss) phases. Figure 4.1c shows a more complex phase diagram between Al and Ti (both metals) exhibiting

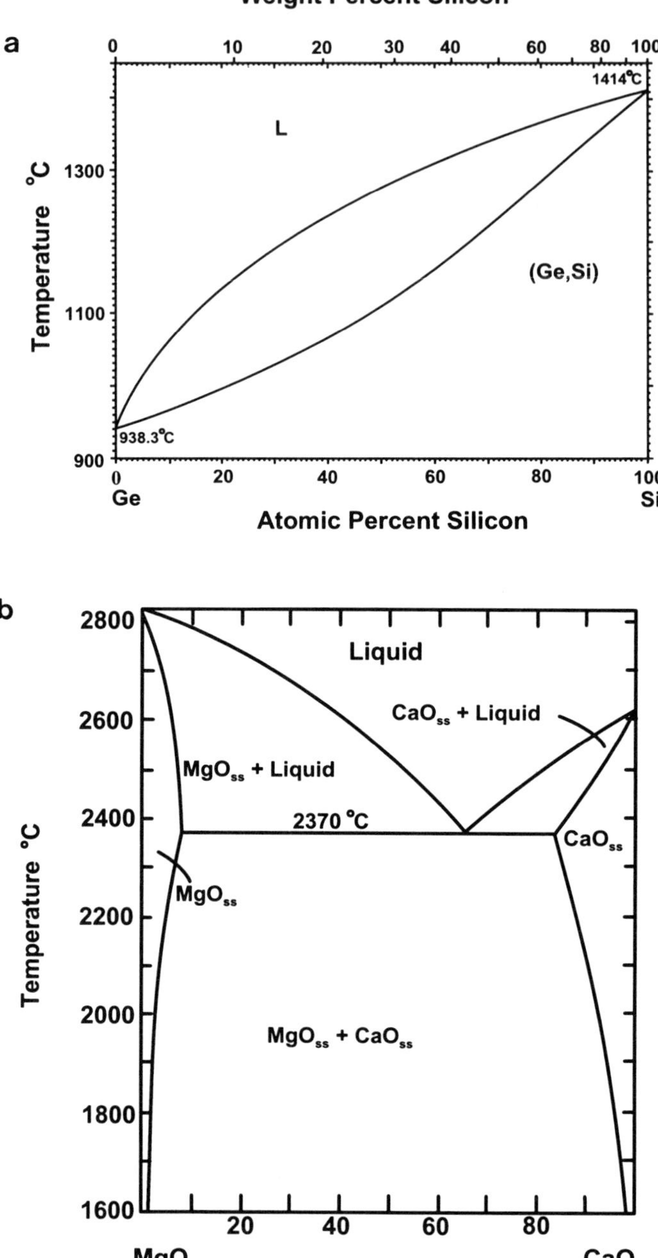

FIG. 4.1. Three typical binary phase diagrams: (a) Ge–Si showing complete solid solubility of Ge in Si and Si in Ge—an isomorphous system (Source: *Binary Alloy Phase Diagrams*, 2nd ed., edited by T. B. Massalski, ASM International, Materials Park, OH, 1990, p. 2001, reprinted with permission); (b) MgO–CaO system showing limited solid solubility of one component in the other and a eutectic reaction (Reprinted with permission of the American Ceramic Society, Post Office Box 6136, Westerville, OH 43086-6136. Copyright 1964 by the American Ceramic Society. All rights reserved)

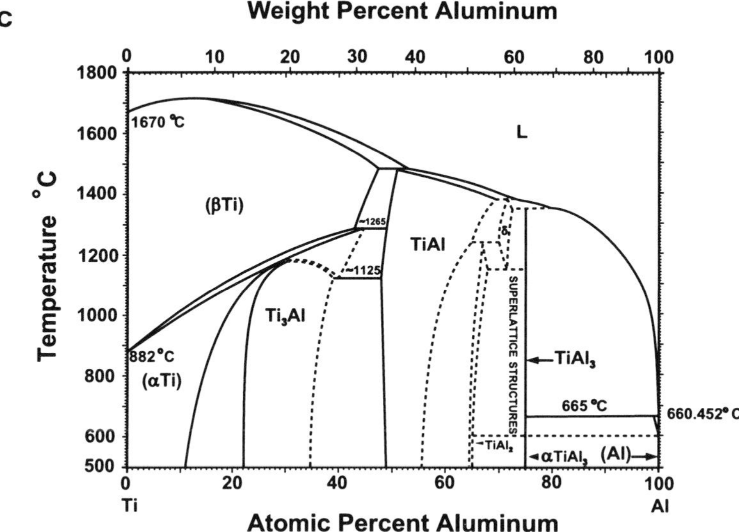

FIG. 4.1. (continued) (c) Ti–Al system exhibiting a number of intermediate phases (Source: *Binary Alloy Phase Diagrams*, 2nd ed., edited by T. B. Massalski, ASM International, Materials Park, OH, 1990, p. 226, reprinted with permission).

a number of intermediate phases, including limited solid solubility of one component in the other. All classes of materials show these (and additional) types of reactions occurring between them. For example, an isomorphous system may be exhibited by metals and ceramics in addition to semiconductors. Similarly, a eutectic system may be found between metals and between semiconductors, and complex phase diagrams are common in all classes of materials. Further, note that phase diagrams are also established for ternary (three components), quaternary (four components), and higher-order systems. However, in this module, we concentrate only on binary (two-component) systems.

The boundary separating a single-phase solid region from a two-phase solid region is called a solvus line, and the composition represented by the solvus line determines the maximum solid solubility of the solute in the solvent at any given temperature. Also note that a horizontal (constant temperature) line drawn across the phase diagram must pass through single-phase and two-phase regions alternately. In other words, two single-phase regions must always be separated by a two-phase region, and two two-phase regions must always be separated by a single-phase region. For example, in Fig. 4.2 notice that the horizontal line representing the temperature T_1 goes from α- to (α + β)- to β-phase fields.

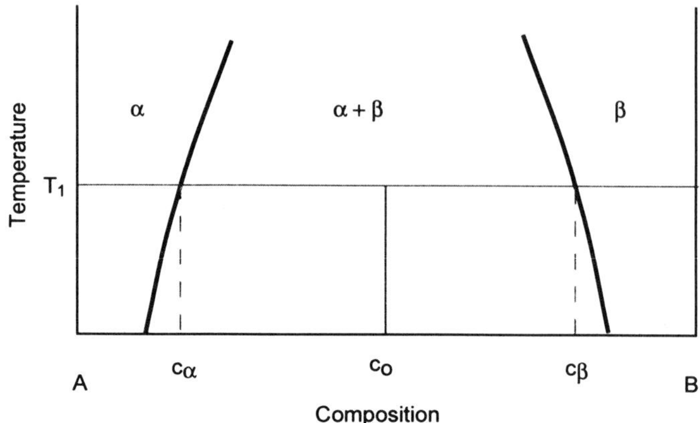

FIG. 4.2 Solid-state portion of a binary phase diagram showing that at any temperature two single-phase regions are always separated by a two-phase field. At any temperature the relative proportions of the two phases in a two-phase field can be found by using the "lever rule."

Determining the position of solvus lines is therefore necessary to establish the maximum solid solubility levels under a given set of conditions. The solvus lines—in fact the full phase diagram—can be determined by several methods, including thermal analysis, dilatometry, and microscopic examination. However, the sensitivity of these methods is limited when solid–solid phase boundaries have to be determined. The x-ray diffraction technique, on the other hand, provides an easy and reliable method for determining the solid–solid phase boundaries. Additionally, x-ray diffraction provides a method to determine the crystal structure of

The relative proportions of the phases in a two-phase mixture can be obtained by using the familiar "lever rule." As an example, let's look at Fig. 4.2 again. If we choose the composition of the alloy to be c_0, then it consists of two phases, α and β. The composition of the α phase, c_α, is given by the point of intersection of the tie line (the horizontal line representing a constant temperature) with the solvus line on the A-rich side. The composition of the coexisting β phase, c_β, is similarly obtained from the point of intersection of the tie line with the solvus line on the B-rich side. The weight fraction of the α and β phases in this alloy at temperature T_1 can be found from the lever rule to be

$$w_\alpha = \frac{c_\beta - c_0}{c_\beta - c_\alpha} \qquad \text{and} \qquad w_\beta = \frac{c_0 - c_\alpha}{c_\beta - c_\alpha}$$

the various phases involved, but it may not be suitable for determining the phase boundaries involving liquid phases.

It is important to remember the following facts while determining the solid–solid phase boundaries by x-ray diffraction techniques.

- In a single-phase region, a change in composition generally produces a change in lattice parameter and, therefore, a shift in the position of the diffraction peaks of that phase. The lattice parameter increases (the diffraction peaks are shifted to lower 2θ values) if the solute has an atomic size larger than that of the solvent. If the solid solution is noncubic (has unequal lattice parameters), then one of the parameters may increase and the other(s) may decrease, but the unit-cell volume always increases. The reverse is true if the atomic size of the solute is smaller than that of the solvent.
- In a two-phase region, a change in composition of the alloy produces a change in the relative amounts of the two phases, but the compositions of the two coexisting phases remain constant. Remember, these compositions may be found as the points of intersection of a "tie line" with the boundaries of the two-phase field. Consequently, there is no change in the lattice parameter(s) and, hence, in the position of the diffraction peaks, but only in the intensity of the reflections. (Estimation of phase proportions from changes in the intensity of reflections is demonstrated in Experimental Module 7.)

Let's now consider how the solid-state portions of a phase diagram can be determined with x-ray diffraction techniques. We first describe the parametric method (in which we use the variation of the lattice parameter with composition) to exactly locate the solvus lines for the α and β phases in Fig. 4.3a. Assume, for simplicity, that the α phase has an fcc structure and the β phase a bcc structure. A series of alloys, 1 to 8, is equilibrated at a temperature T_1, where the α- and β-phase fields are expected to have maximum range. All the alloys are then quenched to room temperature and their x-ray diffraction patterns are recorded. The diffraction patterns of alloys 1 and 2 show the presence of only the α phase, and alloys 7 and 8 show only the β phase. The other alloys show both α and β phases. The lattice parameters of the α phase in alloys 1 to 6 and that of the β phase in alloys 3 to 8 are calculated from the procedures described in Experimental Module 1. The lattice parameters of the α and β phases are plotted in Fig. 4.3b.

Notice that the lattice parameter versus composition curves have two branches: an inclined branch *mn* followed by a horizontal line for the α

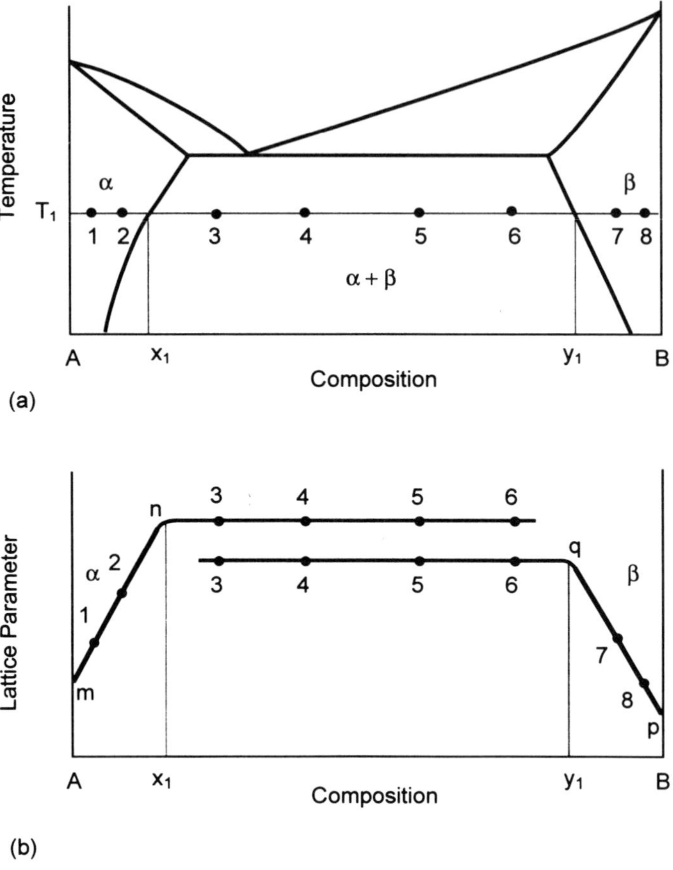

FIG. 4.3. Parametric method to determine the positions of solvus lines in a binary phase diagram. (a) Phase diagram and compositions (positions) of the alloys chosen. Alloys 1 and 2 are in the α-phase field, 7 and 8 in the β-phase field, and 3 through 6 in the two-phase, α + β phase, field. (b) Variation of lattice parameters of the α and β phases with composition.

phase on the A-rich side and an inclined branch *pq* followed by another horizontal line for the β phase on the B-rich side. These two horizontal lines will not meet since the lattice parameters of the saturated α and β solid solutions are different. Further, phases with a "simple" bcc structure normally have a lattice parameter smaller than those with a "simple" fcc structure. The inclined portions at either end of the figure show that the lattice parameter varies with composition in the solid solution region, and the horizontal branches show that the α and β phases in alloys 3 to 6 are saturated, since the lattice parameters do not change with change in alloy composition. The solid solubility of B in A (i.e., the limit of the α-phase field) at T_1 is thus given by the point of intersection (x_1) of the two branches of the lattice parameter curve on the A-rich side. The solid solubility of A in B is similarly given by the point of intersection (y_1) of the two branches of the curve on the B-rich side.

The solid solubility limits of B in A and of A in B at other lower temperatures can be determined in a similar way, but it is not required to use all the alloy compositions; instead, it is sufficient if we use only one two-phase alloy. Let's look at Fig. 4.3a again and concentrate on alloy 3 or 4 (in the two-phase region). The enlarged portion of the A-rich side of the phase diagram is reproduced in Fig. 4.4a, and from this we notice that the solid solubility of B in A decreases with decreasing temperature. Accordingly, the solubility is x_1 at temperature T_1, x_2 at T_2, etc. We now equilibrate alloy 4 at different temperatures, quench it to room temperature, record the x-ray diffraction patterns, and measure the lattice parameter of the α phase. These lattice parameter values are superimposed on the plot we obtained earlier for different alloys at T_1 (Fig. 4.4b). On drawing a horizontal line through this lattice parameter value for an alloy equilibrated at T_2, you will notice that it intersects the *mn* line at a point corresponding to x_2, representing the solid solubility of B in A at this temperature. The solid solubility values at any other temperature can also be found as the point of intersection between the inclined portion *mn* in Fig. 4.4b and the horizontal line drawn through the lattice parameter obtained at that temperature. These maximum solid solubility values at different temperatures can then be plotted in the form of a phase diagram—temperature versus composition—from which the solid solubility at any other temperature can also be determined. The solvus line on the B-rich side of the phase diagram can be established in a similar way.

In reality, the phase diagrams are not quite as simple as we have made them out to be. Quite frequently, they contain intermediate phases—secondary solid solutions or intermetallic compounds (see, for example, Fig. 4.1c). The preceding method can also be used to determine the solvus lines that bound the secondary solid solutions and intermediate phases, but remember that each phase has its own characteristic x-ray diffraction pattern. Also remember that the alloys always need to be equilibrated and that, at any given temperature, irrespective of the number of phases present or complexity of the phase diagram, two single-phase regions are always separated by a two-phase region.

The slope of the lattice parameter versus composition curve (Fig. 4.4b) (determined primarily by the relative sizes of the solute and solvent atoms and to some extent by the strain in the alloy system) is important since it determines the accuracy with which the solvus curve can be determined.

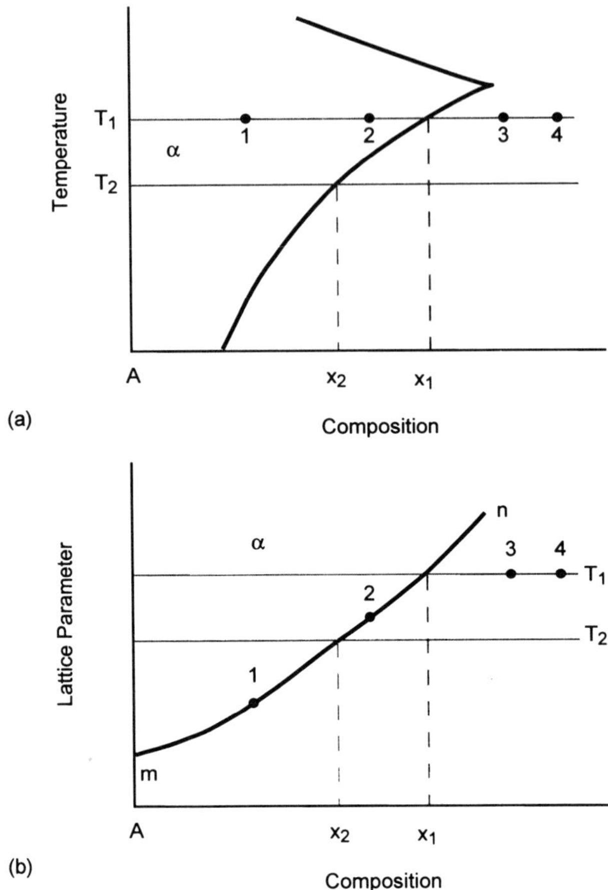

FIG. 4.4. (a) Enlarged portion of the A-rich side of the binary A–B phase diagram in Fig. 4.3a. (b) Variation of lattice parameter of the α phase with B content.

If the curve is nearly flat, i.e., if changes in the composition of the solid solution produce very small changes in the lattice parameter(s), then the composition, as determined from the lattice parameter, is subject to considerable error. Consequently, the location of the solvus line is also in error. However, if the curve is steep, relatively crude lattice parameter measurements may suffice to fix the location of the solvus quite accurately.

The parametric method can be used even when the crystal structure of the phase is so complex that the diffraction pattern cannot be indexed. In that case, the plane spacing corresponding to some high-angle peak, or, even more directly the 2θ value of the peak can be used instead of the lattice parameter, and the procedure described in our example can be followed to determine the solid solubility limits.

The solid-state portion of the full phase diagram can also be determined with this procedure. Let's consider a binary system that exhibits complete solid solubility of one component in the other. Such a system is called *isomorphous* and is formed when the two components involved have

- A small difference in atomic size (usually $<15\%$)
- The same crystal structure
- Similar electronegativities
- The same valency

These empirical rules were postulated by Hume–Rothery and are frequently observed in metallic solid solutions even though there are some exceptions. For example, all the empirical rules are satisfied in the Ag–Cu system, but the alloys exhibit a eutectic reaction and not complete solid solubility of Cu in Ag and of Ag in Cu. It is certain, however, that if

The solvus lines can also be determined by using the disappearing-phase method. In this method, a series of two-phase ($\alpha + \beta$) alloys are equilibrated at a given temperature, quenched to room temperature, and their x-ray diffraction patterns recorded. Since we are now dealing with two-phase alloys, the diffraction pattern consists of two phases, α and β. First we identify the two phases; then we locate one intense peak from each of the phases (the peaks should be close to each other but not overlap) and plot the intensity ratio, I_β/I_α, where I_β and I_α are the relative intensities of the peaks from the β and α phase, respectively, against alloy composition. Since at the $\alpha/(\alpha+\beta)$ solvus line, the amount of β phase is zero (according to the lever rule, the amount of the β phase decreases as you move toward the A-rich side of the phase diagram), extrapolation of the I_β/I_α versus wt% B plot to zero value of I_β/I_α gives the composition of the solvus line at that temperature. The experiment can be repeated at different temperatures, and the solvus composition at different temperatures established.

This method, however, suffers from some inherent disadvantages. First, the plot of I_β/I_α versus wt% B is not linear, and, therefore, a number of experimental points, especially near the phase boundary, are required for high accuracy. Second, the method depends on the disappearance of the β phase (and hence the name disappearing-phase method), so the accuracy depends on the sensitivity of the x-ray diffraction method in detecting small amounts of this phase in a mixture of phases. This sensitivity is very high if the two atoms involved have nearly the same atomic scattering factor (or atomic number) and if the width of the two-phase field is small. Otherwise, the parametric method should be preferred since it depends directly on the lattice parameter measurement of the α phase. In fact, in most cases, the parametric method is more accurate than the disappearing-phase method.

the rules are not obeyed, then the extent of solid solution formation is limited.

In an isomorphous system, the lattice parameter(s) of the solid solution varies continuously, and almost linearly, with composition since in the whole solid solution range the crystal structure is the same but the composition is different. Therefore, if we prepare a series of alloys in this system, record the x-ray diffraction patterns, and calculate the lattice parameter(s), a plot of lattice parameter versus composition (in atomic percent) will show an almost linear variation. This continuous variation of the lattice parameter(s) from one pure component to the other confirms the occurrence of an isomorphous system.

The continuous linear variation of lattice parameter(s) with solute content (in atomic percent) is known as Vegard's law and is commonly obeyed by solid solutions of ionic salts. Vegard's law is rarely, if ever, obeyed by metallic or ceramic solid solutions, which show either a positive or a negative deviation from linearity. Solutions having positive values of heat of mixing show a positive deviation from Vegard's law, whereas solutions having negative values of heat of mixing show negative deviations from Vegard's law. The extent of deviation is especially substantial when the solid solutions have a noncubic structure. This relationship is often used to estimate the approximate solid solubility levels under equilibrium conditions.

Solid Solubility Determination

A materials scientist is very rarely called upon to determine the full phase diagram. More often than not, all that will be required is to calculate or estimate the solid solubility level of one component in another. This can be done very easily and most effectively by x-ray diffraction procedures.

If the identity of the alloy is known and if standard information is available in the literature on the variation of lattice parameter(s) with solute content for this alloy system (see, for example, the Bibliography and Sec. 3.4 of Part I), use these data to generate a master plot of lattice parameter versus composition (at% B). Record the x-ray diffraction pattern of the alloy under consideration, index the diffraction pattern, and calculate the lattice parameter(s), using the methods described in Experimental Module 1, if the material is cubic, or those in Experimental Module 2, if the material is hexagonal. Then estimate the solute content in the alloy from this lattice parameter versus composition master plot.

If lattice parameter(s) versus solute content information for the alloy system under consideration is not available, but you know the lattice

If the measured lattice parameter of the alloy phase is significantly far from the straight-line position, then it is most likely that the alloy under consideration is not in a primary solid solution condition. It may be in a secondary solid solution or intermetallic state. In such a case, Vegard's law plot cannot be used to estimate the solute content in the alloy. (Remember that if you measure the lattice parameter(s) within the secondary solid solution range, then the lattice parameter(s) also varies almost linearly with composition.)

parameter(s) (and crystal structures of the components involved, even though this is not necessary when both the components have the same crystal structure), then you may assume that Vegard's law is obeyed. Plot the lattice parameter(s) of the two components on the y axis and the composition on the x axis, and draw a straight line joining the lattice parameter values of the two components. Record the x-ray diffraction pattern of the alloy whose solid solubility has to be determined, index the diffraction pattern, calculate the lattice parameter(s), and, using your master plot (linear variation of lattice parameter(s) with solute content), estimate the solute content (solid solubility) in the alloy. This does not provide an accurate value of the solute content in the alloy since the measured lattice parameter usually shows either a positive or a negative deviation from the linear plot.

If the crystal structures of the two components are not the same, it is necessary to convert the structure of one component into a hypothetical equivalent structure of the other. An exercise at the end of this module

If the solid solubility level of titanium (hcp crystal structure) in aluminum (fcc crystal structure) has to be determined, then titanium may be assumed to have a hypothetical fcc structure. The lattice parameters of the hexagonal unit cell can be converted to the face-centered cubic unit cell by using the geometrical relationship between the two unit cells. Figure 4.5 shows the atom arrangement on the (100) and (111) planes of the face-centered cubic unit cell and on the (0001) plane of the hexagonal unit cell. Note that the arrangements on $(111)_{fcc}$ and $(0001)_{hcp}$ are identical. Also realize that the atoms are touching along the face diagonal in the face-centered cubic unit cell [see the atom arrangement on the (100) plane] and therefore $a_{fcc} = 2\sqrt{2}r$. In the hexagonal unit cell, the atoms are touching along the a_1 and a_2 axes, and therefore $a_{hcp} = 2r$. From these two relationships, we can write $a_{hcp} = a_{fcc}/\sqrt{2}$. If we assume ideal close packing of atoms in the hexagonal lattice (for which the axial ratio is 1.633), then $c_{hcp} = (2/\sqrt{3})a_{fcc}$.

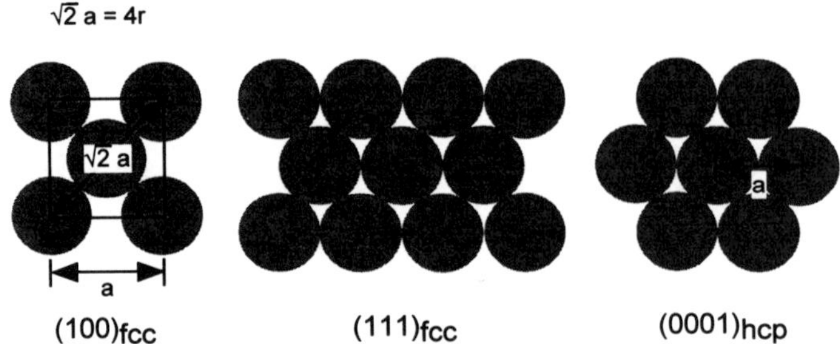

FIG. 4.5. Method to convert the lattice parameters between hexagonal and face-centered cubic lattices.

requires you to use this conversion. Remember that there will be a slight change in the atomic size with a change in the coordination number. For example, if the coordination number decreases from 12 to 8 (i.e., when the structure changes from fcc or hcp to bcc) there is a 3% contraction in atomic size.

Alternatively, the master plot can be established by taking x-ray diffraction patterns of a series of alloys, indexing them, calculating the lattice parameters, and plotting these data versus solute content. You can then accurately determine the solute content in an unknown alloy, using this master plot.

If you do not know the identity (chemical nature, crystal structure, or lattice parameter(s) of the components involved), then you must first identify the substance or determine the crystal structures and lattice parameters of the two components involved, as demonstrated in Experimental Module 8. You may then use the Vegard's law approach to determine the solid solubility levels.

EXPERIMENTAL PROCEDURE

Take fine powders (–325 mesh, <45 μm in size) of MgO and NiO and record their x-ray diffraction patterns, using Cu Kα radiation in the 2θ angular range of 25 to 140°. Thoroughly mix MgO and NiO in the proportions of MgO + 20, 40, 60, and 80 mol% NiO, giving a total of four powder mixtures. Heat-treat these mixtures at 1500°C for 8 h and slowly cool these alloys to room temperature. Record the x-ray diffraction patterns of these reacted mixtures, using Cu Kα radiation. Index all the diffraction patterns by using the procedure in Experimental Module 1 and calculate the lattice parameters. Plot the lattice parameters against mol% NiO and observe whether a continuous linear variation is obtained between lattice parameter and solute content. The x-ray diffraction patterns of MgO, MgO–20NiO, MgO–40NiO, MgO–60NiO, MgO–80NiO, and NiO are presented in Figs. 4.6 to 4.11, in case you are not able to record them yourself.

Note: We are not providing any worked example for this module, since the experiment involves only measurement of the lattice parameters of cubic materials; you did this in Experimental Module 1. If the materials under consideration have a hexagonal structure, their lattice parameters can be determined from the procedures in Experimental Module 2.

RESULTS

Tabulate your calculations for the MgO and NiO powders and all the reacted mixtures in Tables 4.1 to 4.6 and complete the summary table (Table 4.7). Plot the lattice parameters against mol % NiO content, using Fig. 4.12. Does the lattice parameter vary exactly linearly with the NiO content? Comment on this result.

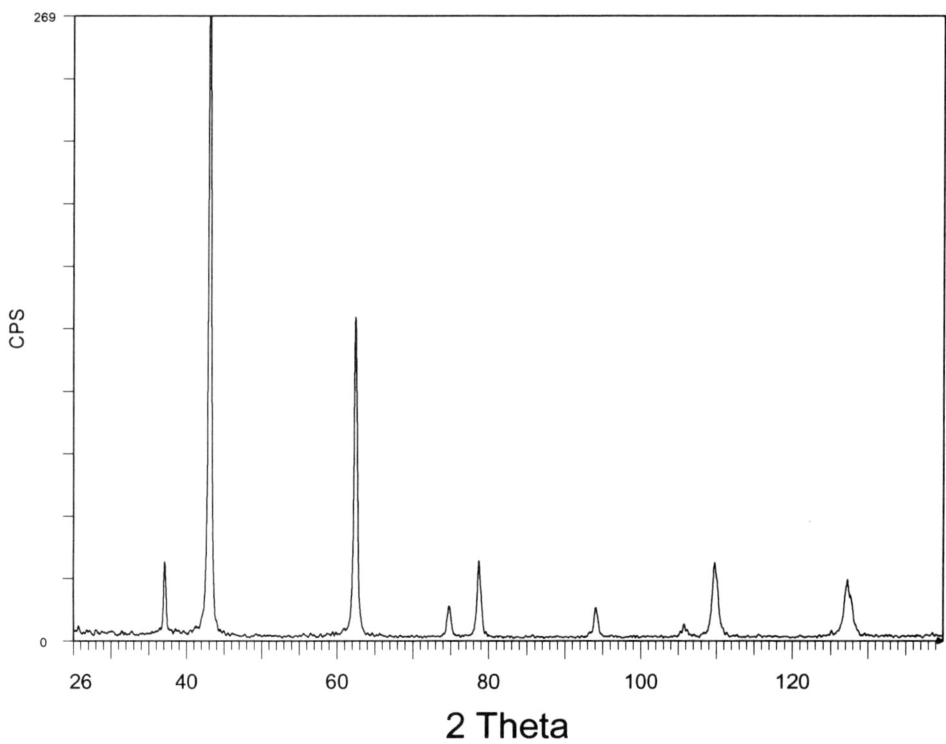

FIG. 4.6 X-ray diffraction pattern of MgO.

TABLE 4.1. Work Table for Pure MgO

Material: Pure MgO Radiation: Cu Kα $\lambda = 0.154056$ nm

Peak #	2θ (°)	$\sin^2\theta$	$\frac{\sin^2\theta}{\sin^2\theta_{min}}$	$\frac{\sin^2\theta}{\sin^2\theta_{min}} \times 3$	$h^2+k^2+l^2$	hkl	a (nm)
1	36.90						
2	42.79						
3	62.12						
4	74.49						
5	78.49						
6	93.93						
7	105.71						
8	109.84						
9	127.16						

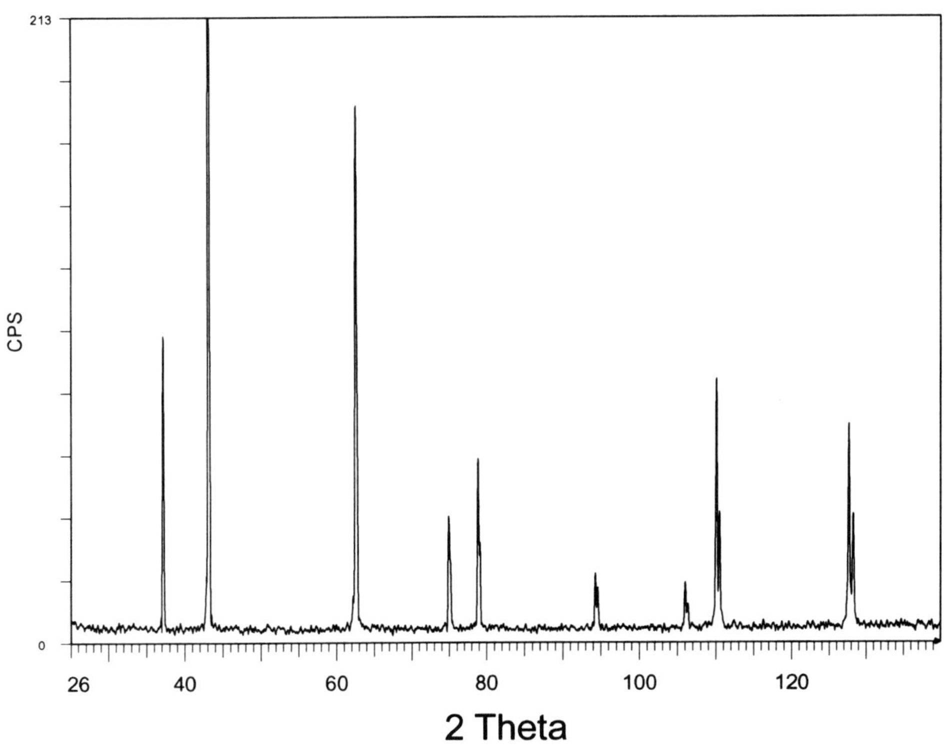

FIG. 4.7. X-ray diffraction pattern of MgO–20 mol% NiO.

TABLE 4.2. Work Table for MgO + 20 mol % NiO

Material: MgO + 20 mol % NiO   Radiation: Cu Kα $\lambda = 0.154056$ nm

Peak #	2θ (°)	$\sin^2\theta$	$\frac{\sin^2\theta}{\sin^2\theta_{min}}$	$\frac{\sin^2\theta}{\sin^2\theta_{min}}\times 3$	$h^2+k^2+l^2$	*hkl*	*a* (nm)
1	36.90						
2	42.91						
3	62.35						
4	74.84						
5	78.73						
6	94.28						
7	105.95						
8	110.07						
9	127.63						

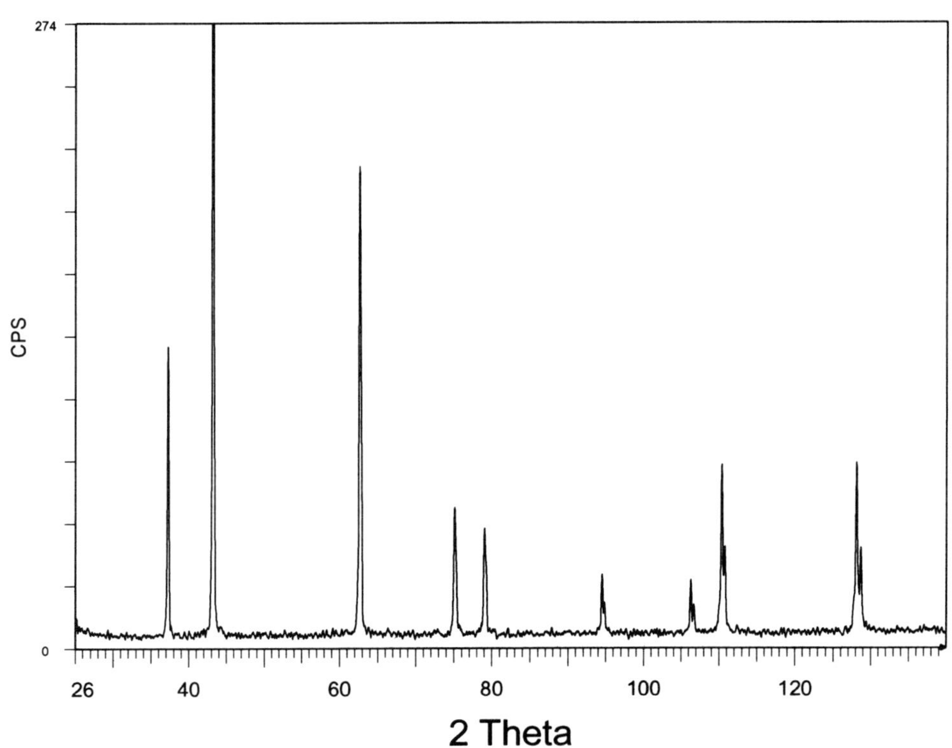

FIG. 4.8. X-ray diffraction pattern of MgO–40 mol% NiO.

TABLE 4.3. Work Table for MgO + 40 mol% NiO

Material: MgO + 40 mol % NiO   Radiation: Cu Kα $\lambda = 0.154056$ nm

Peak #	2θ (°)	$\sin^2\theta$	$\frac{\sin^2\theta}{\sin^2\theta_{min}}$	$\frac{\sin^2\theta}{\sin^2\theta_{min}} \times 3$	$h^2+k^2+l^2$	*hkl*	*a* (nm)
1	37.02						
2	43.03						
3	62.59						
4	74.96						
5	78.97						
6	94.40						
7	106.18						
8	110.19						
9	128.10						

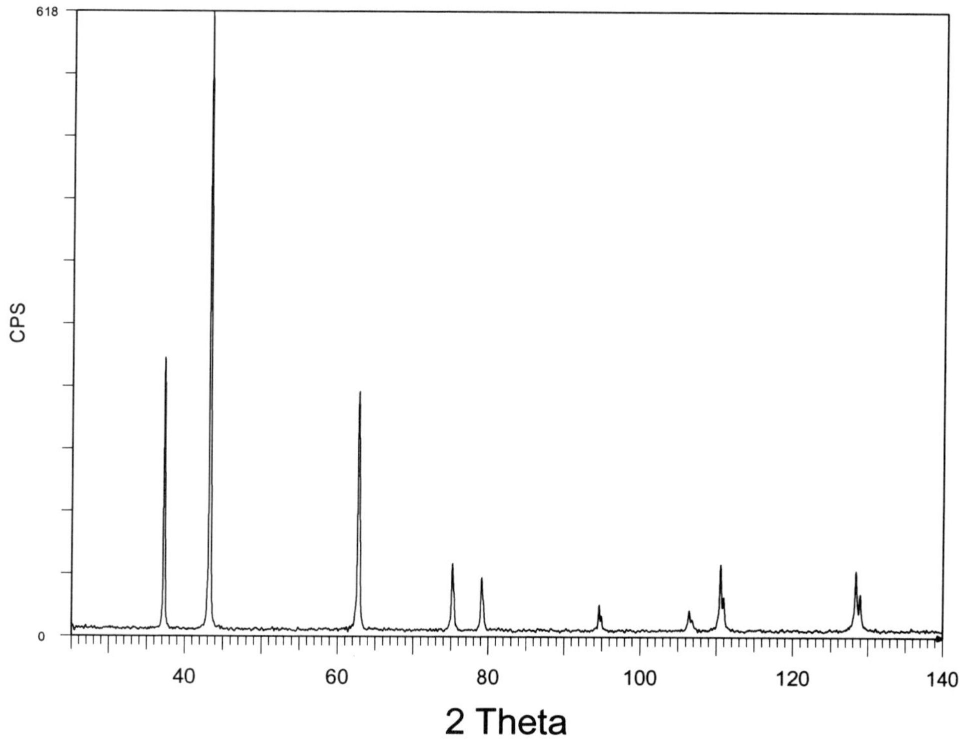

FIG. 4.9. X-ray diffraction pattern of MgO–60 mol% NiO.

TABLE 4.4. Work Table for MgO + 60 mol % NiO

Material: MgO + 60 mol % NiO Radiation: Cu Kα $\lambda = 0.154056$ nm

Peak #	2θ (°)	$\sin^2\theta$	$\frac{\sin^2\theta}{\sin^2\theta_{min}}$	$\frac{\sin^2\theta}{\sin^2\theta_{min}}\times 3$	$h^2+k^2+l^2$	*hkl*	*a* (nm)
1	37.06						
2	43.10						
3	62.72						
4	75.10						
5	79.07						
6	94.60						
7	106.52						
8	110.61						
9	128.48						

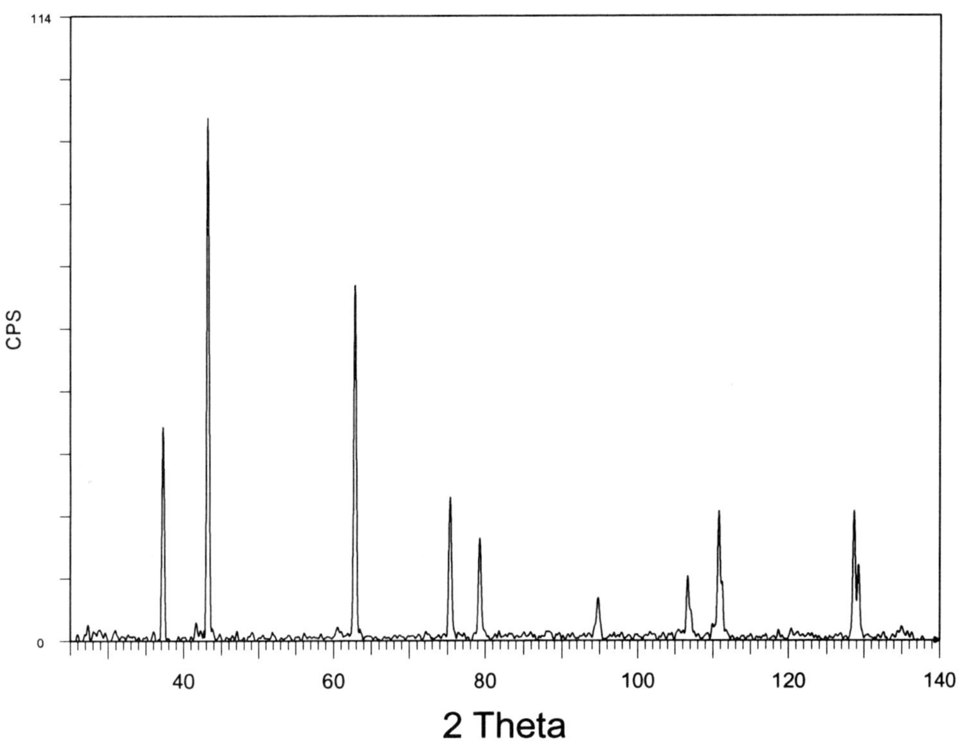

FIG. 4.10. X-ray diffraction pattern of MgO–80 mol% NiO.

TABLE 4.5. Work Table for MgO + 80 mol% NiO

Material: MgO + 80 mol % NiO Radiation: Cu Kα $\lambda = 0.154056$ nm

Peak #	2θ (°)	$\sin^2\theta$	$\frac{\sin^2\theta}{\sin^2\theta_{min}}$	$\frac{\sin^2\theta}{\sin^2\theta_{min}}\times 3$	$h^2+k^2+l^2$	*hkl*	*a* (nm)
1	37.14						
2	43.10						
3	62.72						
4	75.22						
5	79.19						
6	94.84						
7	106.63						
8	110.72						
9	128.82						

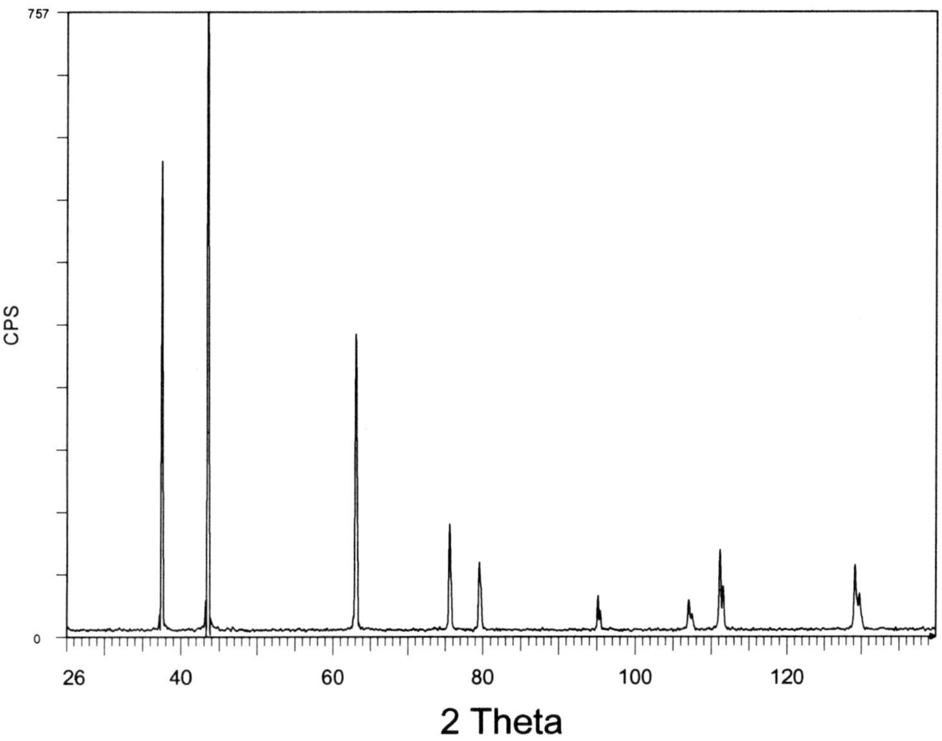

FIG. 4.11. X-ray diffraction pattern of NiO.

TABLE 4.6. Work Table for Pure NiO

Material: Pure NiO　　Radiation: Cu Kα　　$\lambda = 0.154056$ nm

Peak #	2θ (°)	$\sin^2\theta$	$\dfrac{\sin^2\theta}{\sin^2\theta_{min}}$	$\dfrac{\sin^2\theta}{\sin^2\theta_{min}} \times 3$	$h^2+k^2+l^2$	hkl	a (nm)
1	37.30						
2	43.32						
3	62.90						
4	75.43						
5	79.40						
6	95.08						
7	106.99						
8	111.13						
9	129.19						

TABLE 4.7. Summary Table for Experimental Module 4

Alloy #	Composition (mol% NiO)	Lattice parameter (nm)
1	0 (pure MgO)	
2	20	
3	40	
4	60	
5	80	
6	100 (pure NiO)	

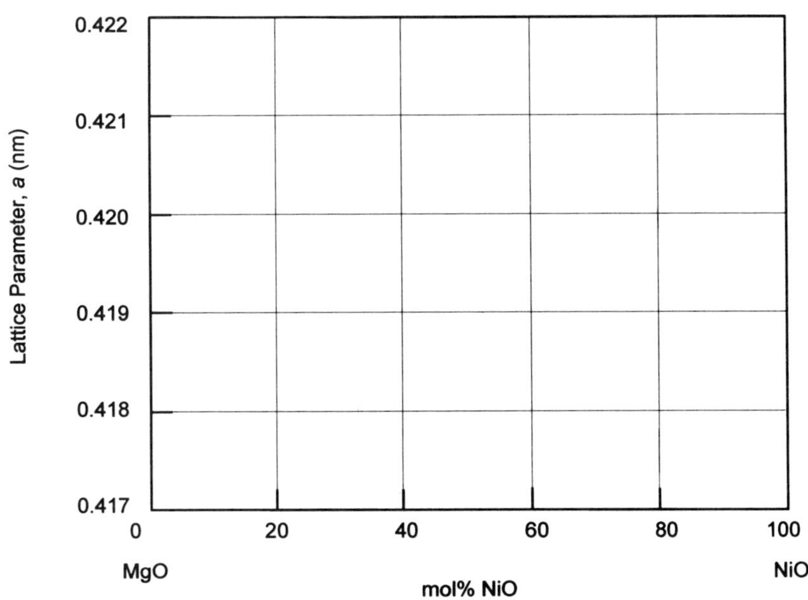

FIG. 4.12. Variation of lattice parameter of MgO–NiO reacted mixtures with NiO content.

EXERCISES

4.1. Draw a schematic phase diagram of a system A–B showing a simple eutectic reaction and substantial solid solubility of B in A (α phase) and negligible solid solubility of A in B (β phase). Sketch the room-temperature x-ray diffraction patterns of several alloys in the system, assuming that the B atom has a smaller size than A. Also assume that both the α and β phases have the fcc structure.

4.2. The solid solubility of titanium (hcp) in aluminum (fcc) can be substantially increased by processing their powders under nonequilibrium conditions, even though the equilibrium solid solubility is virtually nil. Suggest the procedure you would adopt if you need to determine the solid solubility of titanium in aluminum by x-ray diffraction techniques. (Hint: Assume that Vegard's law is obeyed and convert the lattice parameters of the hexagonal unit cell of titanium to a hypothetical face-centered cubic unit cell. The lattice parameters of these metals are $a_{\text{Al}} = 0.4049$ nm, $a_{\text{Ti}} = 0.2951$ nm, and $c_{\text{Ti}} = 0.4683$ nm.)

EXPERIMENTAL MODULE 5

Detection of Long-Range Ordering

OBJECTIVE OF THE EXPERIMENT

To detect the presence of long-range ordering and quantify the long-range order parameter by x-ray diffraction methods.

BACKGROUND AND THEORY

In a binary substitutional solid solution (or compound), the two kinds of atoms (A and B) are arranged more or less at random on the atomic sites of the lattice. When cooled below a critical temperature T_c, the A atoms in the solid solution arrange themselves in an orderly manner on one set of lattice sites, and the B atoms arrange themselves on another lattice site. The material is then said to be in an *ordered* state. When the material is heated above T_c, the atomic arrangement again becomes random and the material is said to be in a *disordered* condition. Since the constituent atoms of the alloy or compound need to occupy specific lattice sites in the ordered condition, the order–disorder transformation usually occurs at compositions in which the two types of atom are in simple ratios, say 1:1, 1:2, 1:3, ... Perfect ordering takes place at the stoichiometric composition. When the order–disorder transformation takes place away from this stoichiometry, then some amount of disordering is present at low temperatures because there is an excess of one type of atoms, and these atoms have to necessarily occupy the "wrong" lattice sites. You should also realize that the order–disorder transformation occurs in ternary and higher-order alloys and in interstitial solid solutions and compounds.

When the periodic arrangement of A and B atoms persists over large distances in the crystal, it is known as long-range order. If all the A and B atoms in the solid solution (or compound) occupy the "right" lattice sites, then ordering is said to be perfect. The degree of perfection is quantified by the long-range order parameter S. For perfect long-range order the value of $S = 1$, and for perfect disorder $S = 0$; it has an intermediate value (between 0 and 1) if some of the atoms occupy the "wrong" lattice sites.

In a binary solid solution A–B, let's assume that the A atoms occupy the α sites and the B atoms the β sites. This situation is considered as the "right" atoms occupying the "right" sites. If an A (or B) atom occupies the β (or α) site, then it is considered that the A (or B) atom is occupying the "wrong" site. With this definition, S can be defined as

$$S = \frac{p - r}{1 - r} \tag{5.1}$$

where p is the fraction of α sites occupied by A atoms (or the β sites occupied by B atoms) and r is the fraction of A (or B) atoms in the alloy. As an example, consider β-brass (an alloy of 50 at% Cu and 50 at% Zn). In the ordered condition, β-brass has the cubic CsCl structure (see Fig. 25 in Part I) in which the Cu atoms occupy the cube corner positions, which have coordinates 0,0,0, and the Zn atoms occupy the body center positions, which have coordinates $\frac{1}{2},\frac{1}{2},\frac{1}{2}$ (or vice versa). Hence, if we consider 100 atoms in a crystal of β-brass, the alloy is perfectly ordered when all 50 Cu atoms occupy the cube corner positions and all 50 Zn atoms occupy the body center positions. Then from Eq. (5.1) we get

$$S = \frac{1 - 0.5}{1 - 0.5} = 1 \tag{5.2}$$

On the other hand, if there is some disordering (e.g., if only 47 Cu atoms occupy the cube corner positions), then $p = 47/50 = 0.94$, and

$$S = \frac{0.94 - 0.5}{1 - 0.5} = 0.88 \tag{5.3}$$

As a consequence, the long-range order parameter S decreases as more and more atoms occupy the "wrong" sites (i.e., the amount of disorder increases), and S reaches a value of zero for complete disorder.

The order–disorder transformation is observed in a variety of alloys and compounds. Most intermetallics and ceramic compounds are in a permanently ordered condition in the solid state; i.e., they do not undergo disordering in the solid state but get disordered only on melting (when they are in the liquid state). On the other hand, some alloys show the order–disorder transformation in the solid state, such as, Cu–Au, Cu–Zn, Fe–Al.

The transformation from the disordered state to the ordered state is accompanied by changes in the physical and mechanical properties of the material. The ordered alloys are usually stronger and harder, have a lower electrical resistivity, and a higher electrical conductivity than their disordered counterparts. Therefore, it is possible to detect the order–disorder transformation by following the changes in these properties as a function of temperature. However, x-ray diffraction is a powerful technique allowing the presence of ordering in a material to be detected and the degree of ordering to be quantified.

We shall use the classic Cu_3Au alloy to study the order–disorder transformation by x-ray diffraction procedures. Since copper and gold atoms satisfy the Hume-Rothery criteria for complete solid solution formation (see Experimental Module 4), Cu–Au alloys form an isomorphous system and the solid (Au,Cu) phase has the fcc structure at all compositions, as shown in Fig. 5.1. At temperatures above T_c ($T_c = 390°C$), the copper and gold atoms in Cu_3Au are arranged at random on the atomic

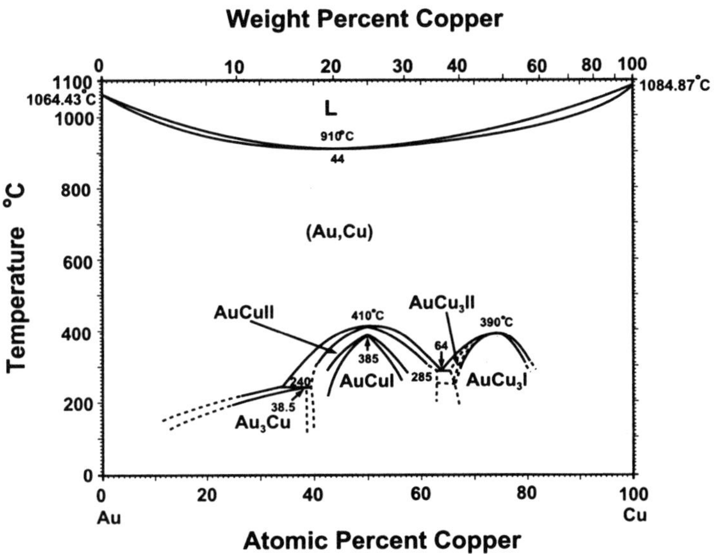

FIG. 5.1. Copper-gold binary phase diagram. Note that the alloy at the composition of 75 at% Cu, i.e., Cu_3Au, undergoes the order–disorder transformation at 390°C. (Source: *Binary Alloy Phase Diagrams,* 2nd ed., edited by T. B. Massalski, ASM International, Materials Park, OH, 1990, p. 360, reprinted with permission).

sites in a face-centered cubic lattice. Since it is not possible to distinguish the copper and gold atoms in the disordered condition, we can consider that each atomic site of the lattice is occupied by an "average" Cu–Au atom made up of $\frac{3}{4}$Cu + $\frac{1}{4}$Au, i.e., the exact proportion of the atoms in the alloy. On cooling the alloy to lower temperatures, below T_c, the constituent atoms occupy specific lattice sites in the original face-centered cubic lattice (ordered condition). Since the ratio of face-centered to cube corner positions in the unit cell is 3:1, the copper atoms occupy the face-centered positions and the gold atoms occupy the cube corner positions. Figure 5.2 shows the unit cells of the compound in both the disordered and ordered conditions. Notice that while the Bravais lattice is face-centered cubic in the disordered condition, it is primitive cubic in the ordered condition. Since the crystal structure in the ordered and disordered conditions is cubic and they have practically the same lattice parameters, there will be little change in the positions of the diffraction peaks, but the specific occupancy of the atoms in the ordered lattice results in additional reflections in the x-ray diffraction pattern (see also the structure factor equations in Appendix 4).

The reason for the additional reflections is due to the arrangement of the atoms on specific lattice sites. For example, if we consider the (100)

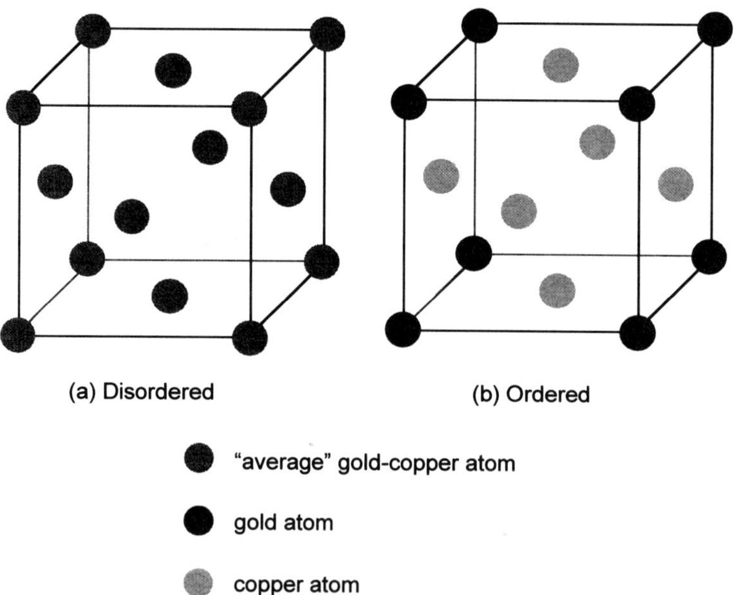

FIG. 5.2. Unit cells of the disordered and ordered structures of Cu_3Au.

planes in the *disordered* condition, the presence of a (200) plane halfway between two (100) planes produces complete cancellation of the 100 reflection. The reason is that all the waves scattered from all the planes have the same amplitude (because all the planes are made up of the "average" Cu–Au atoms, Fig. 5.3a) but are exactly out of phase by $\lambda/2$. However, in the *ordered* condition, even though there is an additional plane of atoms midway between two (100) planes and the scattered waves are also exactly out of phase by $\lambda/2$, the amplitude of the scattered waves is different. The difference in amplitude occurs because the middle plane, the (200) plane, has a different scattering factor since it contains only copper atoms, while the (100) planes contain both copper and gold atoms (Fig. 5.3b). The additional reflections in the ordered condition can also be explained on the basis of structure factor calculations. (We discussed the structure factor in Sec. 2.8 in Part I.) For the disordered solid solution

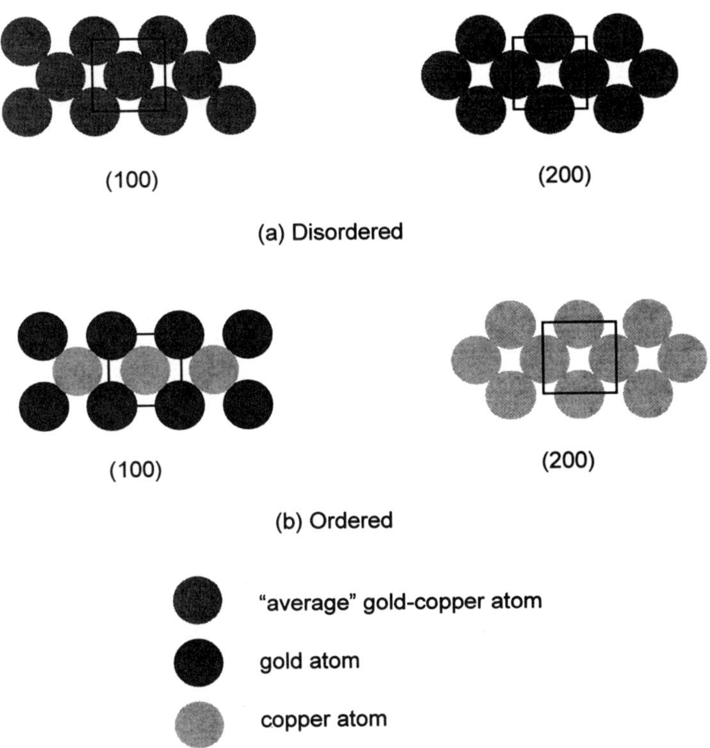

FIG. 5.3. Atomic arrangement on (100) and (200) planes of the Cu_3Au structure in the (a) disordered and (b) ordered conditions.

$$F = 4f_{av} = f_{Au} + 3f_{Cu} \qquad \text{for } h, k, l \text{ unmixed} \tag{5.4}$$

and

$$F = 0 \qquad \text{for } h, k, l \text{ mixed} \tag{5.5}$$

whereas for the completely ordered alloy

$$F = f_{Au} + 3f_{Cu} \qquad \text{for } h, k, l \text{ unmixed} \tag{5.6}$$

and

$$F = f_{Au} - f_{Cu} \qquad \text{for } h, k, l \text{ mixed} \tag{5.7}$$

The ordered alloy thus produces additional reflections in the diffraction pattern for planes with mixed values of *hkl*; i.e., *h, k,* and *l* can be individually either odd or even. These additional reflections are known as *superlattice* reflections. The reflections from planes with *hkl* unmixed are common to both the disordered and ordered lattices and are known as *fundamental* reflections. The fundamental reflections have a much higher intensity than the superlattice reflections, since the structure factor for the former involves the *sum* of the atomic scattering factors while the *difference* is involved for the superlattice reflections. However, note that the intensity of the superlattice reflections in some cases (e.g., when the atomic numbers of the two components involved are very close to each other) might be so small that they may not be visible in the x-ray diffraction patterns. A case in question is the β-brass alloy containing an equal number of Cu and Zn atoms. Since the atomic numbers of Cu and Zn are 29 and 30, respectively, the difference in their atomic scattering factors (which are proportional to the atomic numbers) is very small. Therefore, the intensity of the superlattice reflections is so low that you may not see them in the x-ray diffraction pattern. Whether you actually see them or not, superlattice reflections are direct evidence that ordering has taken place in the alloy.

If the long-range ordering is not perfect (due to some of the gold atoms occupying the face-centered positions or some of the copper atoms occupying the cube corner positions), then the intensity of the superlattice reflections will be lower than when the ordering is perfect. In such a case, the structure factor for the superlattice reflections is

$$F = S(f_{Au} - f_{Cu}) \tag{5.8}$$

Thus, by measuring the intensity I (which is proportional to F^2) of a superlattice reflection relative to that of a closely spaced fundamental reflection, say $(100)_{superlattice}$ and $(111)_{fundamental}$, S can be calculated from the equation

In many alloy systems the order–disorder transformation may involve not only a change in the Bravais lattice (in the same crystal system) but also a change in the crystal system itself. In that case also, additional reflections occur in the diffraction pattern of the ordered alloy. In both cases, whether a change in crystal system is involved or not, it is possible to detect the occurrence of ordering by observing the presence of weak superlattice reflections in the ordered condition.

$$S^2 = \frac{I_{s(dis)}/I_{f(dis)}}{I_{s(ord)}/I_{f(ord)}} \tag{5.9}$$

where I_s and I_f represent the integrated intensities of the superlattice and fundamental reflections, respectively, and the subscripts (dis) and (ord) refer to the disordered and ordered states, respectively.

WORKED EXAMPLE

Let's now see how we can determine the presence of ordering and quantify it in a Cu_3Au alloy. As noted earlier, an alloy of this composition undergoes the order–disorder transformation at 390°C (Fig. 5.1).

A Cu_3Au alloy powder specimen was sealed into *two* glass tubes under vacuum. Both tubes were kept in a furnace maintained at 500°C (above the critical order–disorder transition temperature) for 1 h and then quenched into cold water. This treatment produces the alloy powder in the disordered condition. One tube was retained in the as-quenched (disordered) condition. The remaining tube was held in a furnace maintained at 350°C (below the critical temperature) for 1 h, quenched into cold water, and transferred to another furnace maintained at 280°C and kept there for 90 h. The tube was then slowly cooled to room temperature, inside the furnace, by turning off the power to the furnace. The powder conditions are expected to be as shown in Table 5.1.

TABLE 5.1. Powder Conditions in the Heat-Treated Samples

Powder #	Heat treatment	Powder condition
1	500°C/1 h + water quench	Fully disordered
2	#1 + 350°C/1 h + 280°C/90 h + slow cool to room temperature	Partially ordered

X-ray diffraction patterns of both powders were recorded by using Cu $K\alpha$ radiation in the 2θ angular range of 20 to 140°. The x-ray diffraction patterns, reproduced in Figs. 5.4 and 5.5, were indexed with the procedure in Experimental Module 1. The fundamental and superlattice reflections were identified by looking at the relative intensities and types of the reflections, i.e., whether *h, k,* and *l* were unmixed or mixed. The lattice parameter was calculated in both cases. The relative intensities of all reflections in the partially ordered condition were also noted, taking the intensity of the most intense reflection ($111_{\text{fundamental}}$) as 100%. The long-range order parameter *S* was calculated in the partially ordered condition by using Eq. (5.9). The results are tabulated in Tables 5.2 and 5.3.

The presence of weak additional reflections in powder 2 with *h, k,* and *l* mixed values clearly confirms the presence of ordering in these alloys. Their absence in powder 1 shows that the alloy is truly disordered. The degree of ordering can be quantified by measuring the relative integrated intensities of the superlattice and fundamental reflections in the partially ordered alloy. Table 5.4 lists the relative integrated intensities of the selected reflections. (Remember that the integrated intensity is the area under the peak, and not the peak height.)

The intensity of the superlattice and fundamental reflections in the fully ordered alloy can be theoretically calculated using the expected values of the structure factor, multiplicity factor, Lorentz–polarization factor, etc. These calculations are listed in Table 5.4. The degree of ordering in the partially ordered condition was calculated from Eq. (5.9) to be $S_{\text{partially ordered}} = 0.73$. Even if we had held the powder at 350°C for 1 h and water-quenched it to room temperature (after it was fully disordered), there would have been some ordering. But the longer you heat treat the powder at temperatures below T_c, the higher is the degree of ordering.

The Cu_3Au powder can be heat treated to be in a fully ordered condition, but this takes a long time. Warren (*X-Ray Diffraction*, Addison Wesley, 1969) reports that an annealing treatment for 20 days at 365°C and then for 40 days at 280°C produces complete order in this alloy. Since we may not have such a long time (particularly in a one-semester course), this is not a feasible experiment.

According to theoretical calculations, the intensity of the 100 superlattice reflection in a fully ordered alloy should be 35% of the intensity of the 111 fundamental reflection. However, JCPDS–ICDD (PDF #35-1357) lists the intensity of the 100 reflection as only 17% of that of the 111 reflection. This is probably because the powder for the diffraction pattern

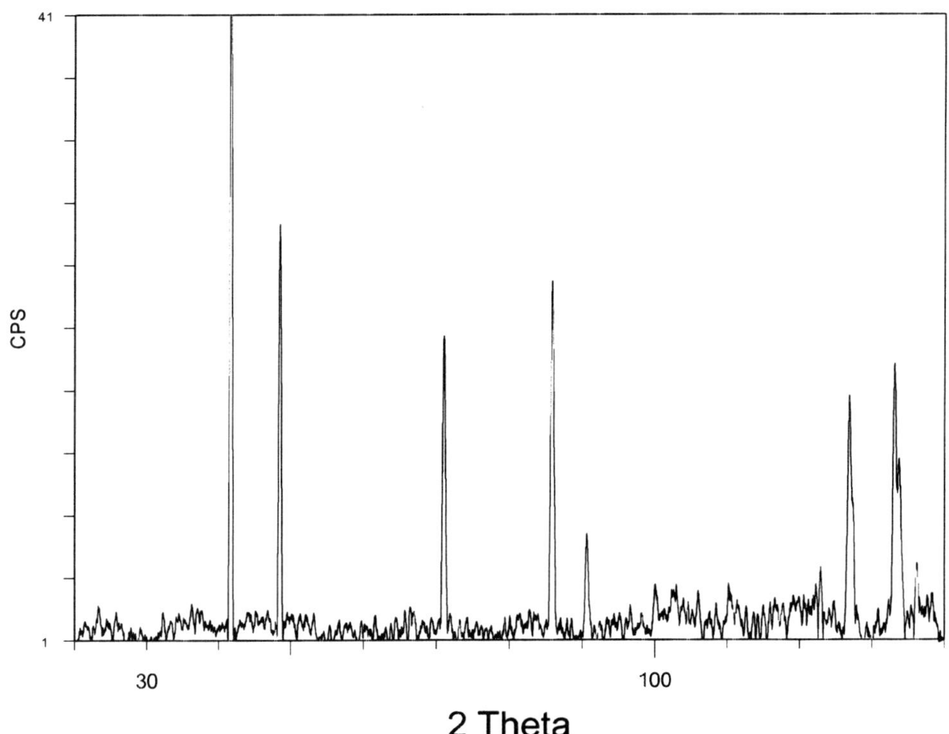

FIG. 5.4. X-ray diffraction pattern of fully disordered Cu_3Au.

TABLE 5.2. Work Table for Fully Disordered Cu_3Au

Material: Fully disordered Cu_3Au			Radiation: Cu Kα				λ = 0.154056 nm
Peak #	2θ (°)	$\sin^2\theta$	$\frac{\sin^2\theta}{\sin^2\theta_{min}}$	$\frac{\sin^2\theta}{\sin^2\theta_{min}}\times 3$	$h^2+k^2+l^2$	*hkl*	*a* (nm)
1	41.53	0.1257	1.000	3.000	3	111	0.3763
2	48.36	0.1678	1.335	4.005	4	200	0.3761
3	70.87	0.3361	2.674	8.021	8	220	0.3758
4	85.71	0.4626	3.680	11.040	11	311	0.3756
5	90.56	0.5049	4.017	12.050	12	222	0.3755
6	110.51	0.6752	5.372	16.115	16	400	0.3750
7	126.90	0.8002	6.366	19.098	19	331	0.3753
8	133.29	0.8428	6.705	20.115	20	420	0.3752

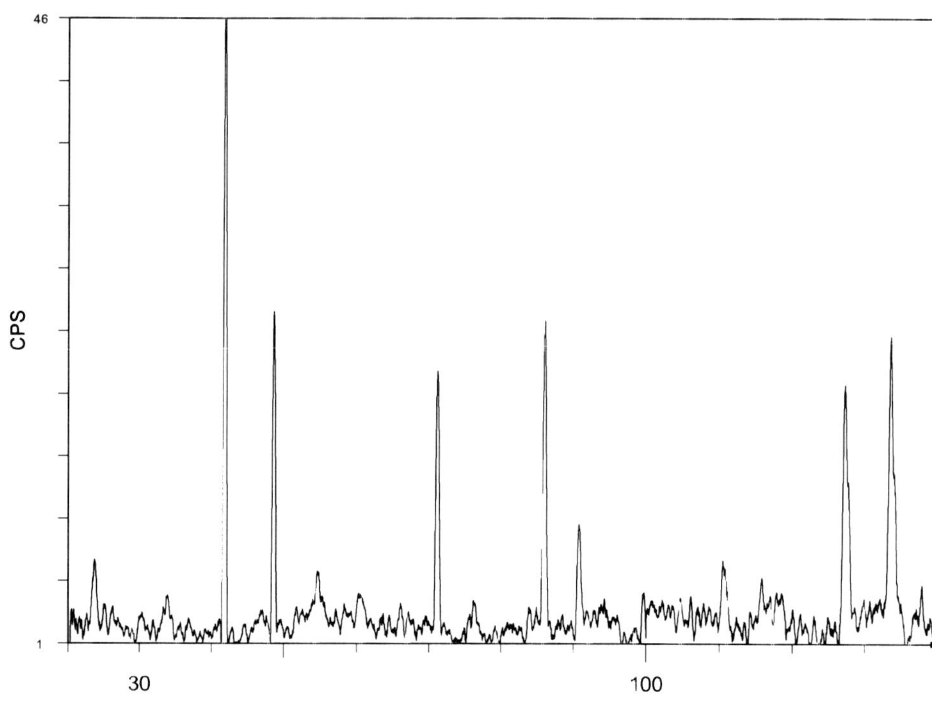

FIG. 5.5. X-ray diffraction pattern of partially ordered Cu_3Au.

TABLE 5.3. Work Table for Partially Ordered Cu_3Au

Material: Partially ordered Cu_3Au			Radiation: Cu $K\alpha$				$\lambda = 0.154056$ nm
Peak #	2θ (°)	$\sin^2\theta$	$\frac{\sin^2\theta}{\sin^2\theta_{min}}$	$\frac{\sin^2\theta}{\sin^2\theta_{min}} \times 1$	$h^2+k^2+l^2$	hkl	a (nm)
1	23.59	0.0418	1.000	1.000	1	100	0.3767
2	33.73	0.0842	2.014	2.014	2	110	0.3754
3	41.65	0.1264	3.024	3.024	3	111	0.3753
4	48.48	0.1686	4.033	4.033	4	200	0.3752
5	54.54	0.2099	5.021	5.021	5	210	0.3759
6	60.35	0.2526	6.043	6.043	6	211	0.3754
7	71.02	0.3374	8.072	8.072	8	220	0.3751
8	76.12	0.3801	9.093	9.093	9	221,300	0.3748
9	85.86	0.4639	11.098	11.098	11	311	0.3751
10	90.67	0.5058	12.100	12.100	12	222	0.3752
11	95.36	0.5467	13.079	13.079	13	320	0.3756
12	100.42	0.5904	14.124	14.124	14	321	0.3751
13	110.51	0.6752	16.153	16.153	16	400	0.3750
14	127.18	0.8022	19.191	19.191	19	331	0.3749
15	133.56	0.8446	20.206	20.206	20	420	0.3748

Note that some of the superlattice reflections, e.g., 310, 410, and 411 (or 330) are so weak that we were not able to measure their peak positions.

TABLE 5.4. Relative Integrated Intensities (in arbitrary units) of Superlattice and Fundamental Reflections in the Heat Treated Cu_3Au Alloys

Alloy	I_{100} (superlattice)	I_{111} (fundamental)
Partially ordered	15.2	81.9
Theoretical values	35	100

in PDF #35-1357 was annealed for only 24 days at 350°C; this might not have produced full ordering!

The results are summarized in Table 5.5.

EXPERIMENTAL PROCEDURE

Very few simple examples of materials exist wherein a clear order–disorder transformation can be observed in the solid state. The number of materials becomes even smaller when we wish to choose a system that retains the cubic structure in the disordered and ordered conditions.

β-Brass (a Cu–50 at% Zn alloy) exists in the ordered condition below between 454 and 468°C (depending on the alloy composition) and in the disordered condition above this temperature. The disordered alloy has a bcc structure, the ordered one a CsCl-type structure. However, this alloy is not suitable for studying the order–disorder transformation by x-ray diffraction techniques. (If you solve Exercise 1 at the end of this module, you will understand why this is so.)

In the Fe–Al system at a composition of 25 at% Al, the alloy exists in the disordered condition (with a bcc structure) at higher temperatures. This phase transforms to the CsCl-type structure (cP2 in the Pearson notation) on cooling to lower temperatures. It is possible to distinguish the disordered and ordered conditions of this Fe–Al alloy by x-ray diffraction techniques, but the high-temperature phase transforms on cooling to the low-temperature cP2 phase so rapidly that to determine the structure of the high-temperature phase you would need access to a high-temperature x-ray diffractometer. A high-temperature x-ray diffrac-

TABLE 5.5. Summary Table for Experimental Module 5

Powder condition	Lattice parameter (nm)	Order parameter, S
Fully disordered	0.3752	0.0
Partially ordered	0.3748	0.7

tometer is not available in most laboratories. (However, the disordered bcc phase can be retained at room temperature either by heavily cold-working the powder or by irradiating the sample with electrons or neutrons.) Further, the disordered bcc phase may also transform to another ordered phase with a different cubic structure (cP16). Hence, this also is not a suitable alloy for our experiments.

The order–disorder transformation can be easily studied in some other copper-base alloys such as Cu– 25 at% Pd or Cu– 25 at% Pt, which show a structural change from the fcc structure (cF4) in the disordered state to the simple cubic structure (cP4) in the ordered state (as in the Cu_3Au alloy). Alternatively, the order–disorder change can be studied in the equiatomic CuAu (Cu–50 at% Au) alloy, which shows a change from the fcc structure in the disordered state to either a tetragonal structure or an orthorhombic structure in the ordered condition (there are two ordered forms). Since it may not be possible to obtain these alloys easily, we are not specifying any material or composition for this experiment.

One possible set of experiments is to heat treat the disordered Cu_3Au alloy powder at 280°C for different periods of time, say 10, 20, 30, 40, 50, and 60 days, and calculate the degree of order for each condition. This will help us understand the kinetics of ordering in this alloy. (Remember to start this experiment at the beginning of the semester!)

Once an alloy composition is chosen, you can do the experiments as described and then identify the presence of ordering (by noting the presence of the weak superlattice reflections in the diffraction pattern) and calculate the order parameter S. Alternatively, you may choose to use the 2θ values listed for the Cu_3Au alloy in the worked example and repeat the calculations and familiarize yourself with the process of identifying the superlattice reflections and estimating the order parameter.

EXERCISES

5.1. β-Brass is an alloy of Cu and Zn with almost 50 at% Zn. The Bravais lattice in the disordered condition is body-centered cubic and that in the ordered condition is primitive cubic (CsCl-type structure). Calculate the relative intensities of the first three superlattice reflections (using Eq. (28) in Part I and ignoring the temperature factor). Assume that the diffraction pattern is recorded with Cu Kα radiation. (*Note*: Superlattice reflections are those for which $h + k + l$ is odd.) Will these reflections be observed in the diffraction pattern? (Hint: Calculate the relative intensities of these reflections with respect to the 110 fundamental reflection whose intensity can be considered as 100%.)

5.2. NaCl has a cubic structure with 4 Na and 4 Cl atoms per unit cell located at

Na	0,0,0	$\frac{1}{2},\frac{1}{2},0$	$\frac{1}{2},0,\frac{1}{2}$	$0,\frac{1}{2},\frac{1}{2}$
Cl	$\frac{1}{2},\frac{1}{2},\frac{1}{2}$	$0,0,\frac{1}{2}$	$0,\frac{1}{2},0$	$\frac{1}{2},0,0$

Calculate the structure factors for the ordered and disordered conditions and identify the superlattice reflections. (Note that NaCl always exists in the ordered condition, but it would be instructive to calculate the structure factor for the disordered condition.)

EXPERIMENTAL MODULE 6

Determination of Crystallite Size and Lattice Strain

OBJECTIVE OF THE EXPERIMENT

To determine the average crystallite size and lattice strain in a powder specimen by x-ray peak broadening analysis.

MATERIALS REQUIRED

Copper metal powder obtained by filing a bulk copper specimen (effectively a cold-worked sample).

BACKGROUND AND THEORY

In deriving Bragg's law, we assume that ideal conditions are maintained during diffraction. These conditions are that the crystal is perfect and that the incident beam is composed of perfectly parallel and strictly monochromatic (single wavelength) radiation. These conditions never actually exist. In fact, only an infinite crystal is really perfect. Finite size alone, of an otherwise perfect crystal, can be considered a crystal imperfection.

The "ideal" size for powder diffraction depends on the relative perfection of the polycrystalline material, but is usually 500 nm (0.5 μm) to 10,000 nm (10 μm). If the crystallites are smaller, the number of parallel planes available is too small for a sharp diffraction maximum to build up, and the peaks in the diffraction pattern become broadened.

A crystal is usually considered perfect when all lattice sites are occupied by atoms and no imperfections (point: vacancies, interstitials; linear: dislocations; planar: grain boundaries, stacking faults; or volume: voids, precipitates) exist in the crystal. Crystals are never perfect in this sense because they contain varying amounts of defects under equilibrium conditions. For example, at room temperature, aluminum has an equilibrium vacancy concentration of 1.4×10^{-13}. The vacancy concentration in semiconductors and ceramics is generally much lower. Although dislocations are not equilibrium defects, they are present in most materials. A dislocation density of about 10^6 m/m^3 is typical for well-annealed metals at room temperature. In single-crystal silicon, made for integrated circuits, the dislocation density is approximately zero. In the present context, a perfect crystal is considered as one in which the interplanar spacings are uniform and there are no distortions in the planes.

Broadening of x-ray diffraction peaks is easily apparent in patterns obtained with a diffractometer, and this information can be directly quantified. However, it is important to realize that broadening of diffraction peaks arises mainly due to three factors.

1. *Instrumental effects*: These effects include imperfect focusing, unresolved α_1 and α_2 peaks, or the finite widths of the α_1 and α_2 peaks in cases where the peaks are resolved. These extraneous sources can cause broadening of the diffraction peaks. Thus, the ideal peak shape—a peak without any noticeable width (Fig. 6.1a)—gets transformed to that shown in Fig. 6.1b due to instrumental effects.

2. *Crystallite size*: The peaks become broader due to the effect of small crystallite sizes, and thus an analysis of peak broadening can be used to determine the crystallite sizes from 100 to 500 nm. Small crystallite sizes introduce additional broadening into the diffraction peaks; therefore, in the presence of instrumental effects, crystallite sizes broaden the peak as in Fig. 6.1c.

3. *Lattice strain*: If all the effects mentioned are simultaneously present in the specimen, the peak will be very broad, as shown in Fig. 6.1d. However, it is possible to separate the individual effects of broadening by following some simple procedures, which will now be described.

Some Definitions

Some terms you will come across in the literature on peak-broadening effects are *domain size*, *crystallite size*, and *grain size*. These terms have created

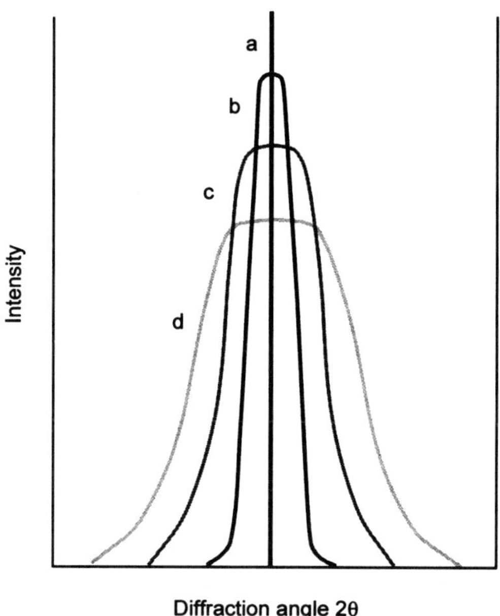

FIG. 6.1. X-ray diffraction peak widths: (a) ideal, (b) due to instrumental effects, (c) superimposition of instrumental and crystallite size effects, and (d) combined effects of instrumental, crystallite size, and lattice strain effects (not drawn to scale).

some confusion, so we will try to clarify their meanings. A domain is that part of the specimen that diffracts the x-ray beam coherently, and it is sometimes called a coherently diffracting domain. The domains form a substructure, which can occur in single crystals and in polycrystalline materials. The individual domains are very slightly misoriented (generally $< 1°$) with respect to each other, and they are smaller than the grain size of the material. For example, the grains may be divided by low-angle grain boundaries (a planar arrangement of individual dislocations). A cold-worked (and partially or fully recovered) metal, for example, has domains (referred to as subgrain boundaries or cells separated by low-angle grain boundaries). A well-annealed metal does not have domains (subgrain boundaries), but does have grains.

For completeness, we mention that faulting in crystals also affects the shape of the diffraction peaks. The effects are complex and depend on the crystal structure and nature of faulting. An excellent treatment of the topic can be found in B. E. Warren (*X-Ray Diffraction*, Addison-Wesley, Reading, MA, 1969, pp. 275–312). These effects will not be considered further in this module.

Recording x-ray diffraction patterns on photographic film in a camera (though not used now) is an ideal way to differentiate qualitatively between materials with coarse- and fine-sized crystallites. The x-ray diffraction patterns are in the form of a series of concentric rings. Each ring represents diffraction from a set of planes in the crystal. When the crystallites are larger than about 10 μm, the number of crystallites in the irradiated portion of the specimen is insufficient to reflect to each portion of the ring, and consequently the ring becomes spotty. Between 500 nm (0.5 μm) and 10,000 nm (10 μm), the number of crystallites is sufficient to produce a smooth and continuous ring pattern, and this is considered the "ideal" size of the crystallites. If the crystallite size is small, say 100 nm (0.1 μm) to 500 nm (0.5 μm), then the diffraction ring becomes broadened. However, if it is extremely small, say << 100 nm (0.1 μm), then the irradiated volume is too small to build up any diffraction ring and diffraction occurs only at low angles. Thus, by observing the diffraction rings in a series of powder patterns made from specimens containing different crystallite sizes, one can qualitatively grade the crystallite sizes in a particular order. This is qualitatively similar to what happens in electron diffraction. The smaller the crystallite size the more broadened are the diffraction rings, and the larger the crystallite size the more spotty the diffraction rings. In fact, one can count the number of spots in a diffraction ring recorded under a given set of conditions (specimen condition, crystallite size, type and radius of camera, operating conditions of the x-ray diffraction unit, etc.) and use such data as standards. The crystallite size of an unknown specimen can then be determined by comparison with these standard patterns (such as ASTM grain size charts).

Crystallite is a more general term and may mean a domain with reference to, for example, a cold-worked (and partially annealed) metal or a grain in the case of a well-annealed metal. In the literature these terms are used interchangeably. In ionic or covalently bonded materials a domain structure is unlikely because the dislocation densities are generally much lower and dislocation movement is more difficult than in a metal. In a nanocrystalline material the grain size is so small (~10 to 100 nm) that it will not be possible to have a substructure, so crystallite, domain, and grain all mean the same thing.

The definitions of the terms we use in this module, peak height and FWHM (full width at half maximum) are indicated in a typical x-ray diffraction peak in Fig. 44 in Chapter 3 of Part I. The broadening is evaluated by measuring the width B, in radians, at an intensity equal to half the maximum intensity (FWHM). Note that B is an angular width, in terms of 2θ (not θ), and not a linear width.

Subtraction of Instrumental Broadening

The individual contributions of small crystallite sizes and lattice strains to the peak broadening can be determined only after first subtracting the effect of instrumental broadening from the experimentally observed peaks. In order to estimate the magnitude of instrumental broadening, it is usual to mix the unknown specimen with some coarse-grained, well-annealed (i.e., strain-free), standard powder whose crystallite size is so large that it does not cause any broadening. For example, silicon powder (a brittle material that can be produced in powder form with no stored lattice strain) with a grain size of about 10 μm is ideal for this purpose.

An x-ray diffraction pattern of the standard powder is recorded under instrumental conditions identical to those of the unknown specimen, so the peak broadening of the standard material due to instrumental effects is exactly the same as instrumental broadening in the diffraction pattern of the unknown specimen. We should take proper care to see that one or more of the x-ray peaks of the standard specimen lie close to (but do not overlap) reflections from the unknown specimen; this ensures that the instrumental broadening is measured at similar Bragg angles for both materials.

Alternatively, we can use a powder of the same composition, if possible, as the experimental material, but in an annealed condition so that the grain size is large and the lattice strain is removed. This procedure has the added advantage that the peak width in the standard and experimental materials is measured at exactly the same angle.

If the observed x-ray peak has a width B_0, and the width due to instrumental effects is B_i, then the remaining width B_r is due to the combined effects of crystallite size and lattice strain:

$$B_r = B_0 - B_i \tag{6.1}$$

This expression is true only when the peak has a Lorentzian (Cauchy) profile. However, if it has a Gaussian profile, a better expression is

$$B_r^2 = B_0^2 - B_i^2 \tag{6.2}$$

In the absence of clear-cut evidence for the exact nature of the peak, we use the geometric mean to get a more nearly correct expression:

$$B_r = \sqrt{(B_0 - B_i)\sqrt{(B_0^2 - B_i^2)}} \tag{6.3}$$

In our analysis, however, we will use Eq. (6.2) to subtract the instrumental broadening from the observed broadening.

Most modern x-ray diffractometers include software for peak profile fitting. This software greatly helps in calculating the intensities, positions, widths, and shapes of the peaks with a far greater precision than is possible with manual measurements or visual inspection of the experimental data.

To apply the proper correction for the instrumental broadening, we must first decide the shape of the peak. The two most commonly assumed line shapes are the Lorentzian (Cauchy) and Gaussian. Even though exact mathematical descriptions can be provided for these two shapes, it is sufficient for us to realize that the most obvious difference between these two is the rate of decay of the tails. Figure 6.2 shows the computer-generated symmetrical Lorentzian (Cauchy) and Gaussian profiles with equal peak heights, 2θ, and FWHM. Note that the tails decay very fast for a Gaussian profile and more slowly for a Lorentzian profile.

Broadening due to Small Crystallite Size

Scherrer has derived an expression for broadening of x-ray diffraction peaks due only to small crystallite sizes:

$$B_{\text{crystallite}} = \frac{k\lambda}{L\cos\theta} \tag{6.4}$$

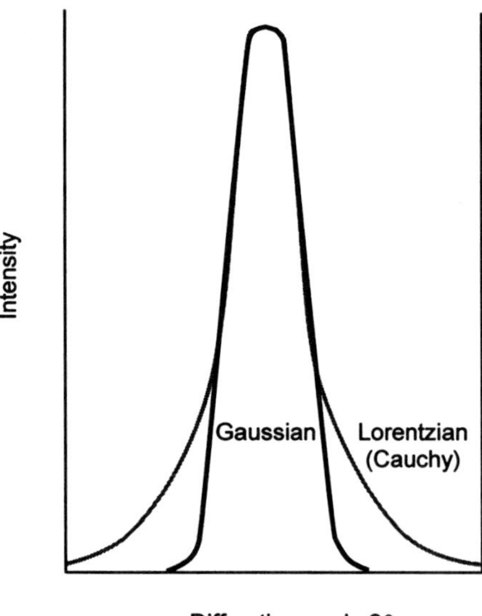

FIG. 6.2. Comparison of symmetrical Lorentzian (Cauchy) and Gaussian x-ray diffraction peak profiles.

where λ is the wavelength of the x-rays used, θ is the Bragg angle, L is the "average" crystallite size measured in a direction perpendicular to the surface of the specimen, and k is a constant. Equation (6.4) is commonly known as the Scherrer equation and was derived based on the assumptions of Gaussian line profiles and small cubic crystals of uniform size (for which $k = 0.94$). However, this equation is now frequently used to estimate the crystallite sizes of both cubic and noncubic materials. The constant k in Eq. (6.4) has been determined to vary between 0.89 and 1.39, but is usually taken as close to unity. Since the precision of crystallite-size analysis by this method is, at best, about ±10%, the assumption that $k = 1.0$ is generally justifiable.

Broadening due to Strain

The lattice strain in the material also causes broadening of the diffraction peaks, which can be represented by the relationship

$$B_{\text{strain}} = \eta \tan\theta \tag{6.5}$$

where η is the strain in the material.

From Eqs. (6.4) and (6.5) it is clear that peak broadening due to crystallite size and lattice strain increases rapidly with increasing θ, but the separation between these two is clearer at smaller θ values, as shown in Fig. 6.3. Since materials may contain both small crystallite sizes and lattice strains, it is desirable to use peaks at smaller diffraction angles to separate these two effects. (Remember that we used high-angle peaks for precise lattice parameter determination.)

The width, B_r, of the diffraction peak after subtracting the instrumental effect can now be considered as the sum of widths due to small crystallite sizes and lattice strains:

$$B_r = B_{\text{crystallite}} + B_{\text{strain}} \tag{6.6}$$

and from Eqs. (6.4) and (6.5) we get

$$B_r = \frac{k\lambda}{L\cos\theta} + \eta\tan\theta \tag{6.7}$$

Multiplying Eq. (6.7) by $\cos\theta$, we get

$$B_r\cos\theta = \frac{k\lambda}{L} + \eta\sin\theta \tag{6.8}$$

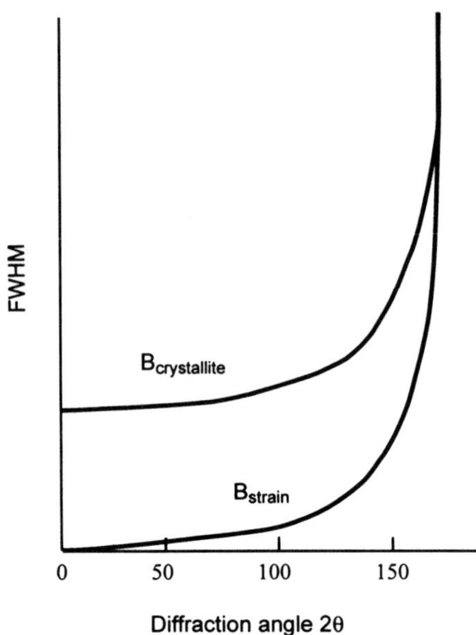

FIG. 6.3. Dependence of FWHM caused by crystallite size and lattice strain on the diffraction angle. Even though both widths increase rapidly with diffraction angle, the separation between them is large at small diffraction angles.

Thus, it is clear that when we plot $B_r \cos\theta$ against $\sin\theta$ we get a straight line with slope η and intercept $k\lambda/L$, as in Fig. 6.4a. The crystallite size L can be calculated from the intercept by using the appropriate values of k (generally taken to be = 1.0) and λ.

Equation (6.8) suggests that the larger the intercept the smaller the size of the crystallites and that for a sufficiently large crystallite size (which does not produce broadening of the diffraction peak), the straight line passes through the origin. The smaller the value of η, i.e., the flatter the straight line, the lower is the amount of strain in the material. These variations are shown schematically in Fig. 6.4b.

Most modern diffractometers have software supplied with the instrument to calculate crystallite sizes and lattice strains in materials. These programs can be used for this purpose. However, we advise you to do the calculations according to the foregoing procedures to be sure you understand them and their physical significance.

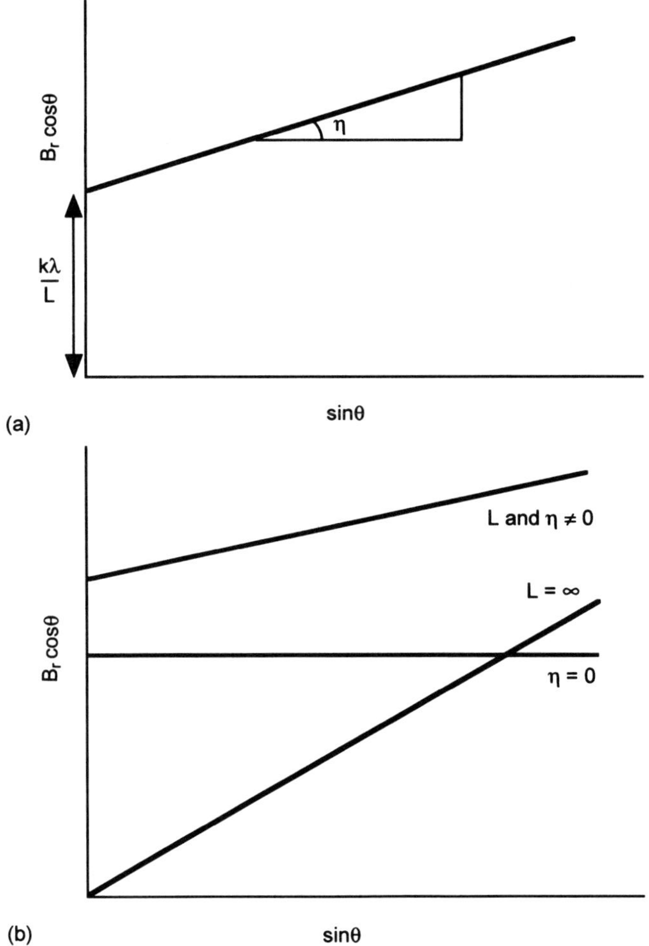

FIG. 6.4. (a) Plot of $B_r \cos\theta$ against $\sin\theta$, indicating that the intercept ($k\lambda/L$) and slope (η) can be used to calculate the crystallite size (L) and lattice strain (η), respectively. (b) Typical plots to show the relative positions of the straight line for very large crystallite sizes ($L = \infty$), no strain ($\eta = 0$), and when both lattice strain and crystallite size contribute to peak broadening.

WORKED EXAMPLE

Let's now work out an example to show how the crystallite size and lattice strain in a material can be calculated from the foregoing procedure. Aluminum metal powder was obtained by filing a bulk specimen; i.e., it is effectively a cold-worked specimen. An x-ray diffraction pattern of this specimen was recorded with Cu Kα radiation in the 2θ angular range of 30 to 70° and is presented in Fig. 6.5. For comparison purposes and to

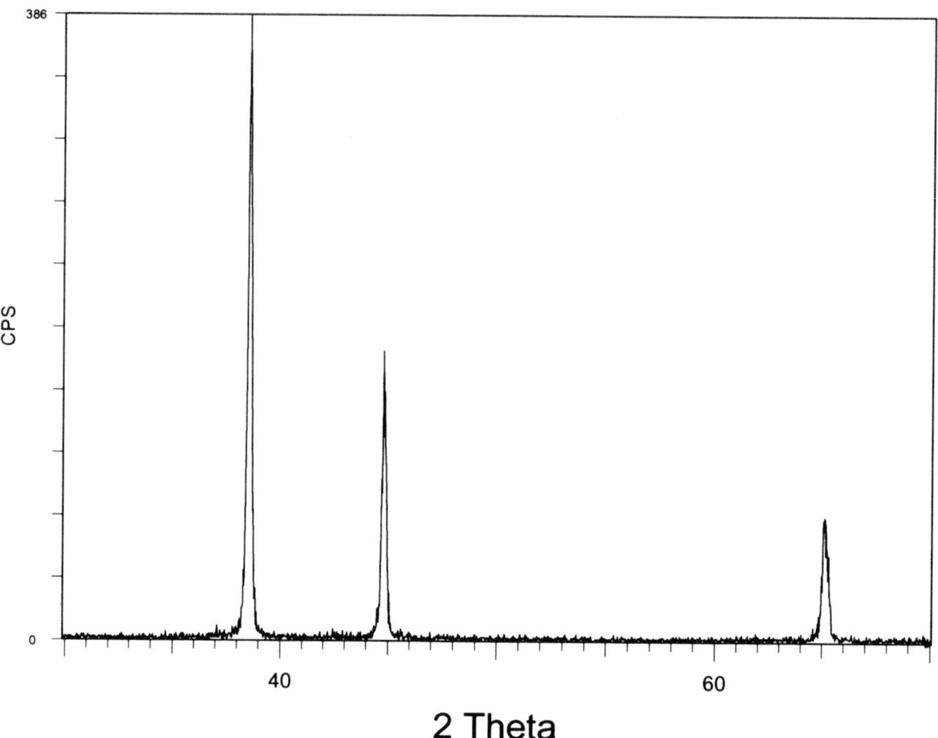

FIG. 6.5. X-ray diffraction pattern of cold-worked aluminum.

calculate the instrumental broadening, the diffraction pattern of an annealed aluminum specimen is presented in Fig. 6.6. Note that there are three peaks in this angular range. In Experimental Module 1, this pattern was indexed, and we found that these three reflections have indices 111, 200, and 220 at the 2θ values of 38.52, 44.76, and 65.13°, respectively. From the x-ray diffraction pattern of annealed aluminum (Fig. 6.6), the instrumental broadening B_i was calculated as the FWHM for the three

TABLE 6.1. Full-Width at Half-Maxima of Annealed Aluminum Specimen

Material: Annealed aluminum		Radiation: Cu Kα		λ = 0.154056 nm
Peak #	2θ (°)	*hkl*	FWHM (°)	FWHM (rad) = B_i
1	38.52	111	0.103	1.8×10^{-3}
2	44.76	200	0.066	1.2×10^{-3}
3	65.13	220	0.089	1.6×10^{-3}

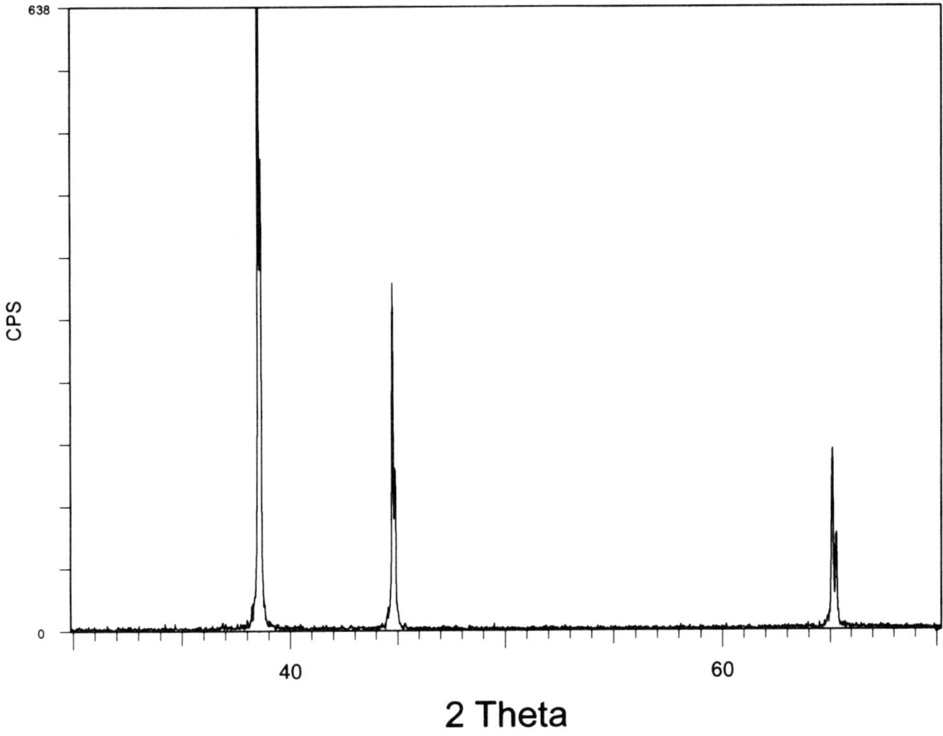

FIG. 6.6. X-ray diffraction pattern of annealed aluminum.

reflections. These calculations (measuring the width of the reflections B_0) were repeated for the experimental pattern recorded from the cold-worked aluminum sample. The remaining broadening was obtained by subtracting the instrumental broadening B_i from the observed broadening B_0, using Eq. (6.2). The results are tabulated in Tables 6.1 and 6.2.

A graph is now plotted between $B_r \cos\theta$ on the y axis and $\sin\theta$ on the x axis (Fig. 6.7). From the slope of this straight line the strain is calculated as 3.5×10^{-3}, and from the intercept the crystallite size is calculated as 90 nm.

TABLE 6.2. Calculations for Cold-Worked Aluminum Specimen

Material: Cold-worked aluminum			Radiation: Cu Kα			λ = 0.154056 nm	
Peak #	2θ (°)	sin θ	*hkl*	B_0 (°)	B_0 (rad)	$B_r^2 = B_0^2 - B_i^2$	$B_r \cos\theta$
1	38.51	0.3298	111	0.187	3.3×10^{-3}	2.8×10^{-3}	2.6×10^{-3}
2	44.77	0.3808	200	0.206	3.6×10^{-3}	3.4×10^{-3}	3.1×10^{-3}
3	65.15	0.5384	220	0.271	4.7×10^{-3}	4.4×10^{-3}	3.7×10^{-3}

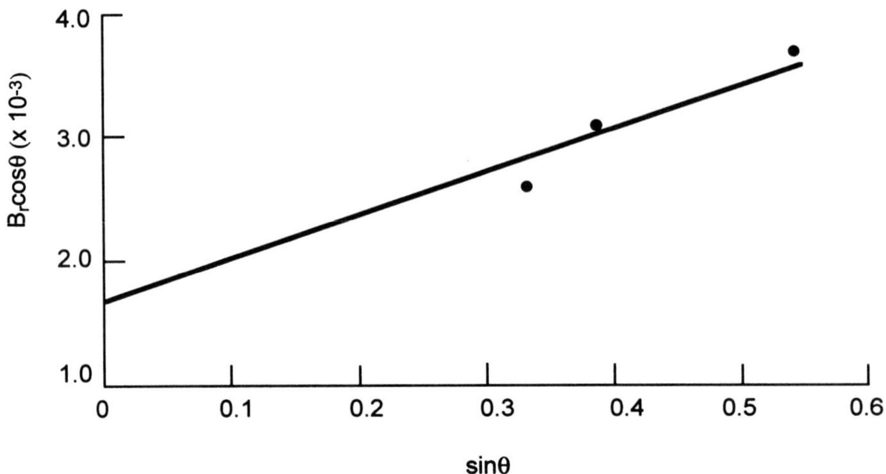

FIG. 6.7. Plot of $B_r \cos\theta$ versus $\sin\theta$ for cold-worked aluminum.

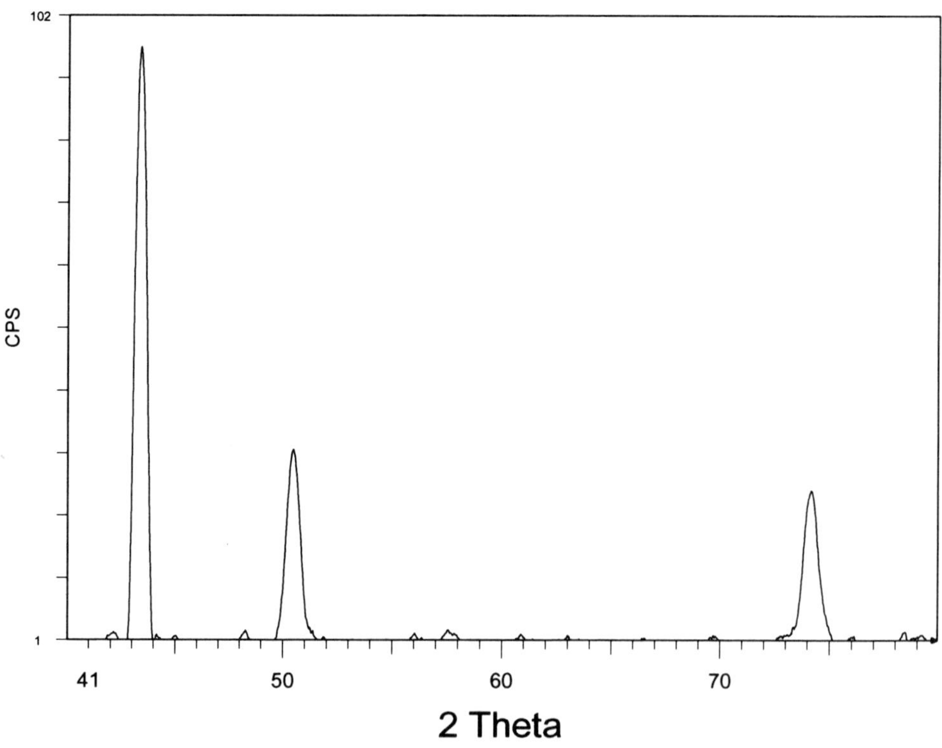

FIG. 6.8. X-ray diffraction pattern of cold-worked copper.

EXPERIMENTAL PROCEDURE

Record the x-ray diffraction pattern from a heavily-cold-worked copper powder, using Cu Kα radiation in the angular range of 2θ from 40 to 80°. Identify the *hkl* values of the different peaks based on your earlier indexing of this pattern in Experimental Module 1. Measure the FWHM for all the peaks. (Remember that the widths are in 2θ angular measure and you should express them in radians.) Evaluate the peak broadening due to instrumental effects, using a well-annealed copper specimen. Subtract this instrumental peak width from the observed peak widths of copper, and determine the crystallite size and lattice strain from all the reflections in the pattern. We have provided x-ray diffraction patterns of a heavily-cold-worked copper powder (Fig. 6.8) and a well-annealed copper powder (Fig. 6.9) if you are not able to record the diffraction patterns yourself.

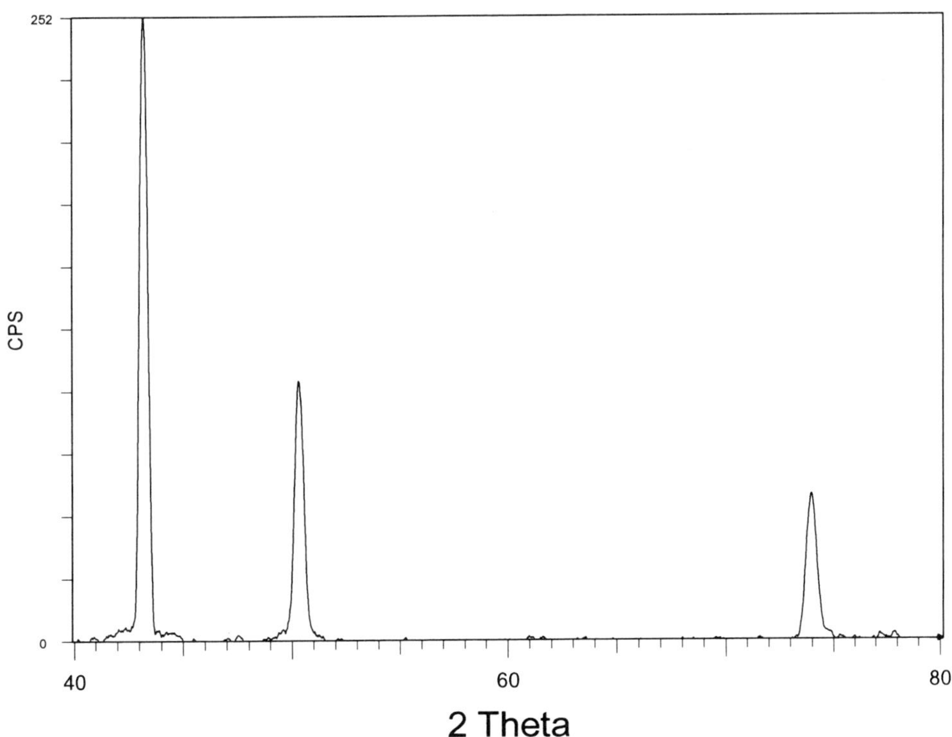

FIG. 6.9. X-ray diffraction pattern of annealed copper.

TABLE 6.3. Work Table for Annealed Copper Specimen

Material: Annealed copper Radiation: Cu Kα $\lambda = 0.154056$ nm

Peak #	2θ (°)	*hkl*	FWHM (°)	FWHM (rad) = B_i
1	43.16	111	0.260	
2	50.30	200	0.330	
3	73.99	220	0.438	

TABLE 6.4. Work Table for Cold-Worked Copper Specimen

Material: Cold-worked copper Radiation: Cu Kα $\lambda = 0.154056$ nm

Peak #	2θ (°)	sin θ	*hkl*	B_0 (°)	B_0 (rad)	$B_r^2 = B_0^2 - B_i^2$	$B_r \cos\theta$
1	43.31		111	0.465			
2	50.40		200	0.541			
3	74.19		220	0.676			

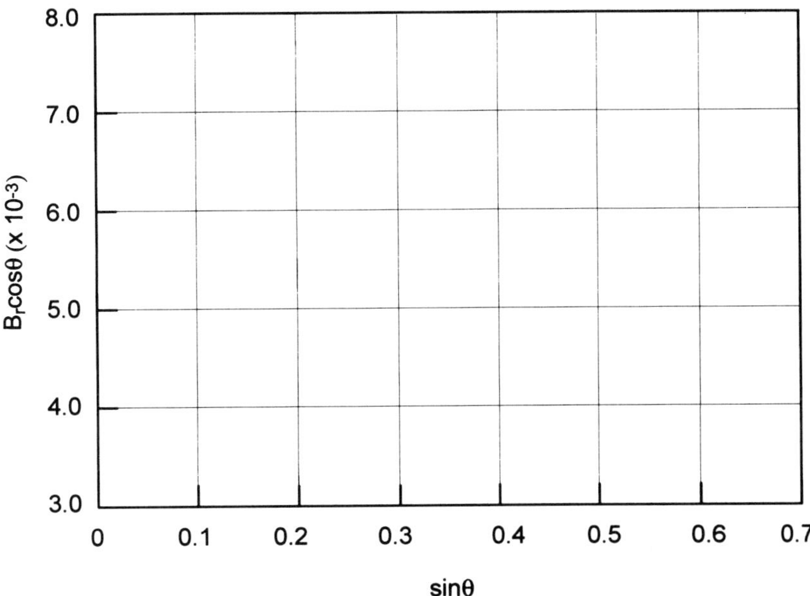

FIG. 6.10. Plot of $B_r \cos\theta$ versus $\sin\theta$ for cold-worked copper.

RESULTS

Tabulate your measurements in Table 6.3.

Further calculations can be done as shown in Table 6.4.

Now plot a graph with $B_r \cos\theta$ on the y axis and $\sin\theta$ on the x axis (Fig. 6.10). The slope of this straight line will be the strain (η), and the intercept is $k\lambda/L$, from which the crystallite size (L) can be estimated. Thus, in the cold-worked copper sample, the crystallite size is _____nm, and the strain is _____.

EXERCISE

6.1. Assume that x-ray diffraction patterns were recorded from an MgO specimen containing crystallites with a diameter of 100, 80, 60, or 40 nm. Calculate the width B (in degrees 2θ) due to the small crystal size alone, assuming that $\theta = 40°$ and $\lambda = 0.154$ nm. For 60-nm-diameter particles, calculate B for $\theta = 20$, 40, and 60°.

EXPERIMENTAL MODULE 7

Quantitative Analysis of Powder Mixtures

OBJECTIVE OF THE EXPERIMENT

To determine the amount of a crystalline phase present in a multiphase material.

MATERIALS REQUIRED

A mixture of thoroughly mixed MgO and CaO powders.

BACKGROUND AND THEORY

Chemical analysis by x-ray diffraction techniques depends on the fact that a phase produces a characteristic x-ray diffraction pattern (a "fingerprint") whether the phase is present alone or in combination with another phase. If the material under examination is a pure metal, its lattice parameter(s) can be measured from the x-ray diffraction pattern and its identity evaluated by referring to a reference book (see the Bibliography). On the other hand, if it is a single-phase alloy (e.g., a solid solution or an intermediate phase), its lattice parameter(s) can be measured and the amount of solute present (the chemical composition) can be determined from a standard lattice parameter versus composition plot. Such methods are useful, for example, in determining the instantaneous solute concentration during diffusion studies using two dissimilar metals. Often, a materials scientist will be called upon to determine quantitatively the amount of one phase present in a mixture of phases, and we describe methods of doing this determination in this module.

The intensity of the diffraction peak of a particular phase in a mixture of phases depends on the concentration of that phase in the mixture. However, the relation between intensity and concentration is not linear because the diffracted intensity depends markedly on the absorption coefficient of the mixture, which itself varies with the concentration of the phase.

The intensity of the diffracted beam from a single-phase polycrystalline powder specimen containing randomly oriented grains in the form of a flat plate in a diffractometer is

$$I = \left(\frac{I_0 A \lambda^3}{32\pi r}\right)\left[\left(\frac{\mu_0}{4\pi}\right)^2 \frac{e^4}{m^2}\right]\frac{1}{v^2}\left[F^2 p\left(\frac{1+\cos^2 2\theta}{\sin^2\theta\cos\theta}\right)\right]\frac{e^{-2M}}{2\mu} \tag{7.1}$$

where I is the integrated intensity of the diffraction peak, I_0 is the incident beam intensity, A is the incident beam cross-sectional area, λ is the incident beam wavelength, r is the radius of the diffractometer circle, μ_0 is a constant with a value of $4\pi \times 10^{-7}$ m kg C^{-2}, e is the electron charge, m is the electron mass, v is the unit-cell volume, F is the structure factor, p is the multiplicity factor, θ is the Bragg angle, e^{-2M} is the temperature factor (M is commonly expressed as $M = B\,((\sin\theta)/\lambda)^2$, where the factor

---

The mass absorption coefficient, μ/ρ (μ is the linear absorption coefficient and ρ is the density), of a mixture of phases is the weighted average of the mass absorption coefficients of its constituents and is given by the relation

$$\left(\frac{\mu}{\rho}\right)_{\text{mixture}} = w_1\left(\frac{\mu}{\rho}\right)_1 + w_2\left(\frac{\mu}{\rho}\right)_2 + \ldots$$

where w_1, w_2, . . . are the weight fractions of phases 1, 2, . . . and $(\mu/\rho)_1$, $(\mu/\rho)_2$, . . . are the mass absorption coefficients of phases 1, 2, ... This relationship is valid whether the substance under consideration is a mechanical mixture, a solid solution, or a chemical compound, and whether it is in the solid, liquid, or gaseous state. The linear absorption coefficient μ, on the other hand, depends on the physical and chemical states of the material. The values of μ/ρ depend on the wavelength of the x-rays used; μ/ρ values of different elements for Cu Kα radiation are listed in Appendix 5.

Quantitative information about the temperature factor is not easy to obtain, and it is very difficult to determine it very accurately. The factor M involves mean-square displacements of atoms, u (whose calculations are extremely difficult), and the Debye characteristic temperature of the specimen. Therefore, in our calculations we ignore the effect of the temperature factor. The error caused by this is not serious because we are using the x-ray diffraction patterns recorded at room temperature and because we are taking ratios of integrated intensities of two closely spaced low-angle reflections.

B contains the temperature term), and μ is the linear absorption coefficient. Recall that $(1 + \cos^2 2\theta)/(\sin^2 \theta \cos \theta)$ is the Lorentz–polarization factor. Equation (7.1) is true for a pure element or a single-phase alloy. If we wish to calculate the intensity of a reflection from one phase (α) in a mixture of phases (α, β, . . .), then we multiply the right-hand side of Eq. (7.1) by c_α the volume fraction of the α phase in the mixture. The intensity of a reflection of the α phase can then be written as

$$I_\alpha = \frac{K_1 c_\alpha}{\mu_m} \tag{7.2}$$

where K_1 is a constant. Because we are now dealing with a mixture, we can replace the linear absorption coefficient μ by μ_m, the linear absorption coefficient for the mixture.

Equation (7.2) can be expressed in a more useful form by writing μ_{m} in terms of concentration or weight fraction. For a binary mixture containing two phases α and β, the mass absorption coefficient is expressed as

$$\frac{\mu_m}{\rho_m} = w_\alpha \left(\frac{\mu_\alpha}{\rho_\alpha}\right) + w_\beta \left(\frac{\mu_\beta}{\rho_\beta}\right) \tag{7.3}$$

where w represents the weight fraction, μ the linear absorption coefficient, and ρ the density. The weight of a unit volume of the mixture is ρ_m, and the weight of α in the mixture is $w_\alpha \rho_m$. Therefore, the volume of α in the mixture is $w_\alpha \rho_m/\rho_\alpha$, which is equal to c_α. A similar expression can be written for c_β. Hence, we can rewrite Eq. (7.2) as

$$I_\alpha = \frac{K_1}{\mu_m} \frac{w_\alpha}{\rho_\alpha} \rho_m$$

or

$$I_\alpha = \frac{K_1\, w_\alpha}{\rho_\alpha(\mu/\rho)_m} \tag{7.4}$$

Even though we do not know the value of K_1, it will cancel when we take the ratio of I_α to the intensity of some standard reference peak. The value of w_α or c_α can then be found from this ratio.

In practice, c_α can be determined by considering the reference peak in three different ways:

1. External standard method (a peak from pure α)
2. Direct comparison method (a peak from another phase in the mixture)
3. Internal standard method (a peak from a foreign material mixed with the specimen)

In all these methods the linear absorption coefficient μ_m of the mixture is strongly dependent on c_α and can have a large effect on the value of I_α. The second method is the most useful in studying phase proportions in materials in a laboratory, and therefore we use this method to show how the phase proportions can be calculated in a two-phase alloy. The third method is frequently used in an industrial environment because of its simplicity. We now describe each method in some detail.

External Standard Method

For the pure α phase Eq. (7.2) can be rewritten as

$$I_{\alpha P} = \frac{K_1}{\mu_\alpha} \tag{7.5}$$

where the subscript P denotes diffraction from the pure phase. Dividing Eq. (7.4) by Eq. (7.5) eliminates the unknown constant K_1, and we get

$$\frac{I_\alpha}{I_{\alpha P}} = \frac{K_1\, w_\alpha}{\rho_\alpha\,(\mu/\rho)_m}\,\frac{\mu_\alpha}{K_1}$$

or

$$\frac{I_\alpha}{I_{\alpha P}} = \frac{w_\alpha\,(\mu/\rho)_\alpha}{(\mu/\rho)_m} \tag{7.6}$$

Knowing the mass absorption coefficients of each of the two phases and the intensity ratio $I_\alpha/I_{\alpha P}$, we can calculate the value of w_α from Eq. (7.6). If the mass absorption coefficients of the individual phases are not known, a calibration curve can be prepared by measuring $I_\alpha/I_{\alpha P}$ for mixtures of known composition. From these $I_\alpha/I_{\alpha P}$ versus w_α plots, we can calculate w_α for a given value of $I_\alpha/I_{\alpha P}$.

Here are some difficulties with this method:

- A specimen of pure α must be available as a reference material; this may not always be possible, especially when we are dealing with metastable phases.
- Measurements of I_α and $I_{\alpha P}$ must be made under identical conditions.
- The variation of the intensity ratio $I_\alpha/I_{\alpha P}$ with w_α is not generally linear; it is linear only when the mass absorption coefficients of the two phases involved are nearly equal to each other. Otherwise there will be a positive (when the β phase has a smaller mass absorption coefficient) or a negative deviation (when β has a larger mass absorption coefficient) (see Exercise 7.2).

Direct Comparison Method

A sample of the pure phase is not required in this method because the reference peak is from another phase in the mixture. This method has been frequently used to estimate phase proportions in a mixture of phases, e.g., the amount of retained austenite in a quenched steel specimen, or the amount of different iron oxides in the oxide scale on steel, or the amount of any one phase in a mixture of phases.

In Eq. (7.1) put

$$K_2 = \frac{I_0 A \lambda^3}{32\pi r}\left[\left(\frac{\mu_0}{4\pi}\right)^2 \frac{e^4}{m^2}\right] \tag{7.7}$$

and

$$R = \frac{1}{v^2}\left[F^2 p\left(\frac{1+\cos^2 2\theta}{\sin^2\theta\cos\theta}\right)\right]e^{-2M} \tag{7.8}$$

The diffracted intensity is then

$$I = \frac{K_2 R}{2\mu} \tag{7.9}$$

where K_2 is a constant, independent of the type and amount of the diffracting substance but dependent only on the diffractometer. On the other hand, R, depends on the nature of the phase and θ and *hkl*. Equation (7.9) can be written for the two different phases α and β as

$$I_\alpha = \frac{K_2 R_\alpha c_\alpha}{2\mu_m} \tag{7.10}$$

and

$$I_\beta = \frac{K_2 R_\beta c_\beta}{2\mu_m} \tag{7.11}$$

By dividing Eq. (7.10) by Eq. (7.11) we eliminate K_2 and μ_m and get

$$\frac{I_\alpha}{I_\beta} = \frac{R_\alpha c_\alpha}{R_\beta c_\beta} \tag{7.12}$$

The value of c_α / c_β can then be obtained from a measurement of I_α / I_β and a calculation of R_α and R_β, which requires a knowledge of the crystal structures and lattice parameters of the two phases. Once the ratio of c_α / c_β is found, the value of c_α (or c_β) can be obtained from the additional relationship

$$c_\alpha + c_\beta = 1 \tag{7.13}$$

This method can be extended to a situation when the sample contains more than two phases.

Some of the precautions you should take in using this procedure are:

- The α and β reflections chosen should not overlap or be very close to each other (when it is difficult to calculate the individual intensities) but should be well separated from each other.
- The volume of the unit cell should be calculated from the observed lattice parameters (since we may be dealing with solid solutions and not pure components).
- In measuring the diffracted intensity, you should measure the integrated intensity and not the peak height because peak height may be different due to broadening depending upon the crystallite size and lattice strain in the sample (see Experimental Module 6), but the integrated intensity will be the same.

Internal Standard Method

In the internal standard method a diffraction peak from the phase (whose volume fraction is being determined) is compared with a peak from a standard substance mixed with the specimen in known proportions. This method is therefore restricted to powders.

Suppose we wish to determine the amount of phase A in a mixture of phases A, B, C, and D, where the relative proportions of the other phases (B, C, and D) may vary from specimen to specimen. We mix a known amount of a standard substance S to a known amount of the original specimen to form a new mixture. If the volume fraction of phase A is c_A in the original specimen and c'_A in the new mixture, and c_S is the volume fraction of the standard in the new mixture, then Eq. (7.2) for a particular reflection from phase A can be written as

$$I_A = \frac{K_3\, c'_A}{\mu_m} \tag{7.14}$$

and that from the standard as

$$I_s = \frac{K_4\, c_s}{\mu_m} \tag{7.15}$$

where K_3 and K_4 are new constants. Dividing Eq. (7.14) by Eq. (7.15) gives

$$\frac{I_A}{I_S} = \frac{K_3\, c'_A}{K_4 c_S} \tag{7.16}$$

The volume fraction of phase A in the new mixture can now be written for a number of components as

$$c'_A = \frac{w'_A/\rho_A}{w'_A/\rho_A + w'_B/\rho_B + w'_C/\rho_C + w'_D/\rho_D + w_S/\rho_S} \tag{7.17}$$

and a similar expression exists for c_S. Then

$$\frac{c'_A}{c_S} = \frac{w'_A\, \rho_S}{\rho_A\, w_S} \tag{7.18}$$

Substituting of Eq. (7.18) into Eq. (7.16) gives

$$\frac{I_A}{I_S} = \frac{K_3}{K_4} \frac{\rho_s}{\rho_A\, w_S} w'_A \tag{7.19}$$

Since K_3, K_4, ρ_S and ρ_A are all constants, and if w_S is kept constant in all the composite specimens, then Eq. (7.19) can be rewritten as

$$\frac{I_A}{I_S} = K_5\, w'_A \tag{7.20}$$

where K_5 is another constant. The weight fractions of A in the original (w_A) and the composite (w'_A) are related to each other as

$$w'_A = w_A(1 - w_S) \tag{7.21}$$

Since w_S is a constant, from Eqs. (7.20) and (7.21), we have

$$\frac{I_A}{I_S} = K_6\, w_A \tag{7.22}$$

Therefore, the intensity ratio of a peak from phase A and a peak from the standard S varies linearly with the weight fraction of phase A in the original specimen. By mixing known concentrations of A and a constant concentration of a suitable standard (for example, fluorite has been found to be a suitable standard in measuring the quartz content of industrial dusts) and measuring the ratio of I_A/I_S, we can establish a calibration curve, from which the concentration of A in an unknown specimen can be determined.

WORKED EXAMPLE

Let's now apply the direct comparison method to calculate the volume fraction of the silicon phase in an aluminum–silicon powder mixture. Small quantities of fine aluminum and silicon powders (–325 mesh, < 45 μm in size) were thoroughly mixed, and an x-ray diffraction pattern was recorded in the 2θ angular range of 25 to 50° with Cu Kα radiation. The pattern (Fig. 7.1) was indexed by using the procedure in Experimental Module 1. The pairs of $(111)_{Al}$ and $(111)_{Si}$ and $(200)_{Al}$ and $(220)_{Si}$ reflections were considered to estimate the amount of the silicon phase present in the mixture. The results are tabulated in Tables 7.1 and 7.2.

Integrated intensity of the $(111)_{Al}$ peak, I_{Al} = 231.0
R value for $(111)_{Al}$ peak, R_{Al} = 36,747,529
Integrated intensity of the $(200)_{Al}$ peak, I_{Al} = 110.6
R value for $(200)_{Al}$ peak, R_{Al} = 17,905,944
Integrated intensity of the $(111)_{Si}$ peak, I_{Si} = 195.2
R value for $(111)_{Si}$ peak, R_{Si} = 34,603,534
Integrated intensity of the $(220)_{Si}$ peak, I_{Si} = 110.4
R value for $(220)_{Si}$ peak, R_{Si} = 23,078,484

Since $I_{Al}/I_{Si} = R_{Al}\, c_{Al}/R_{Si}\, c_{Si}$, the equation can be rewritten as

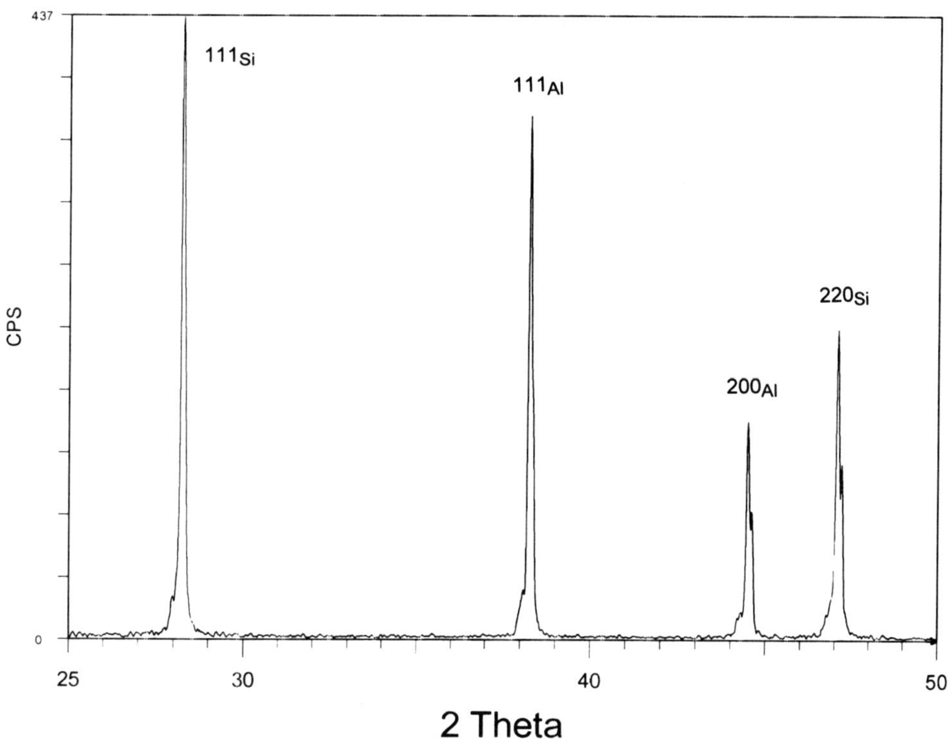

FIG. 7.1. X-ray diffraction pattern of an Al–Si powder mixture.

$$\frac{c_{\mathrm{Al}}}{c_{\mathrm{Si}}} = \frac{I_{\mathrm{Al}}}{I_{\mathrm{Si}}} \frac{R_{\mathrm{Si}}}{R_{\mathrm{Al}}}$$

Therefore, for $(111)_{\mathrm{Al}}$ and $(111)_{\mathrm{Si}}$,

$$\frac{c_{\mathrm{Al}}}{c_{\mathrm{Si}}} = \frac{231}{195.2} \times \frac{34603534}{36747529} = 1.114$$

and for $(200)_{\mathrm{Al}}$ and $(220)_{\mathrm{Si}}$,

$$\frac{c_{\mathrm{Al}}}{c_{\mathrm{Si}}} = \frac{110.6}{110.4} \times \frac{23078484}{17905944} = 1.291$$

Further, since $c_{\mathrm{Al}} + c_{\mathrm{Si}} = 1$, $c_{\mathrm{Si}} = 0.473$ for the $(111)_{\mathrm{Al}}$ and $(111)_{\mathrm{Si}}$ combination and 0.436 for the $(200)_{\mathrm{Al}}$ and $(220)_{\mathrm{Si}}$ combination. Thus, the average volume fraction of silicon in the mixture is 0.45. Consequently, the volume fraction of aluminum is 0.55.

TABLE 7.1. Calculation of R_{Al} for the Al Phase in an Al–Si Powder Mixture

Reflection	111	200
2θ (°)	38.52	44.76
Lattice parameter (nm), *a*	0.4049	0.4049
Volume of unit cell (nm³), *v*	0.06638	0.06638
$\frac{\sin\theta}{\lambda}$	2.141	2.471
Atomic scattering factor, *f* (from Appendix 3)	8.98	8.56
F^2 (use the appropriate equations from Appendix 4)	1290	1172
Multiplicity factor, *p* (from Appendix 6)	8	6
Lorentz–polarization factor $\left(\frac{1+\cos^2 2\theta}{\sin^2\theta\cos\theta}\right)$ (from Appendix 7)	15.69	11.22
$R = \frac{1}{v^2}\left[F^2 p\left(\frac{1+\cos^2 2\theta}{\sin^2\theta\cos\theta}\right)\right]e^{-2M}$ [a]	36,747,529	17,905,944

[a]We are ignoring the temperature factor e^{-2M} since the diffraction pattern is recorded at room temperature, we are using low-angle reflections, and because we are taking the ratios of *R* values.

TABLE 7.2. Calculation of R_{Si} for the Si Phase in an Al–Si Powder Mixture

Reflection	111	220
2θ (°)	28.41	47.35
Lattice parameter (nm), *a*	0.5431	0.5431
Volume of unit cell (nm³), *v*	0.16019	0.16019
$\frac{\sin\theta}{\lambda}$	1.593	2.607
Atomic scattering factor, *f* (from Appendix 3)	10.68	8.79
F^2 (use the appropriate equations from Appendix 4)	3650	4945
Multiplicity factor, *p* (from Appendix 6)	8	12
Lorentz–polarization factor $\left(\frac{1+\cos^2 2\theta}{\sin^2\theta\cos\theta}\right)$ (from Appendix 7)	30.41	9.98
$R = \frac{1}{v^2}\left[F^2 p\left(\frac{1+\cos^2 2\theta}{\sin^2\theta\cos\theta}\right)\right]e^{-2M}$ [a]	34,603,534	23,078,484

[a]We are ignoring the temperature factor e^{-2M} since the diffraction pattern is recorded at room temperature, we are using low-angle reflections, and because we are taking the ratios of *R* values.

EXPERIMENTAL PROCEDURE

Take a thoroughly mixed powder mixture of MgO and CaO and record the x-ray diffraction pattern, using Cu Kα radiation in the 2θ angular range of 25 to 70° (Fig. 7.2). You will have diffraction peaks from both the MgO and CaO phases in your pattern. Since both of them have a cubic structure, index the diffraction peaks, using the procedure in Experimental Module 1, and calculate the lattice parameter of each of the two phases. Choose two suitable pairs of reflections from the two phases (for example, the 200 and 220 reflections of MgO and 111 and 220 reflections of CaO) and calculate the proportion of the MgO and CaO phases, using the direct comparison method. Even though we can take the pairs of 111 and 200 reflections of the two phases, since both of them have the fcc structure, the $(111)_{MgO}$ and $(200)_{CaO}$ peaks are situated very close to each other. In fact, they almost overlap, consequently the intensities add up. Therefore,

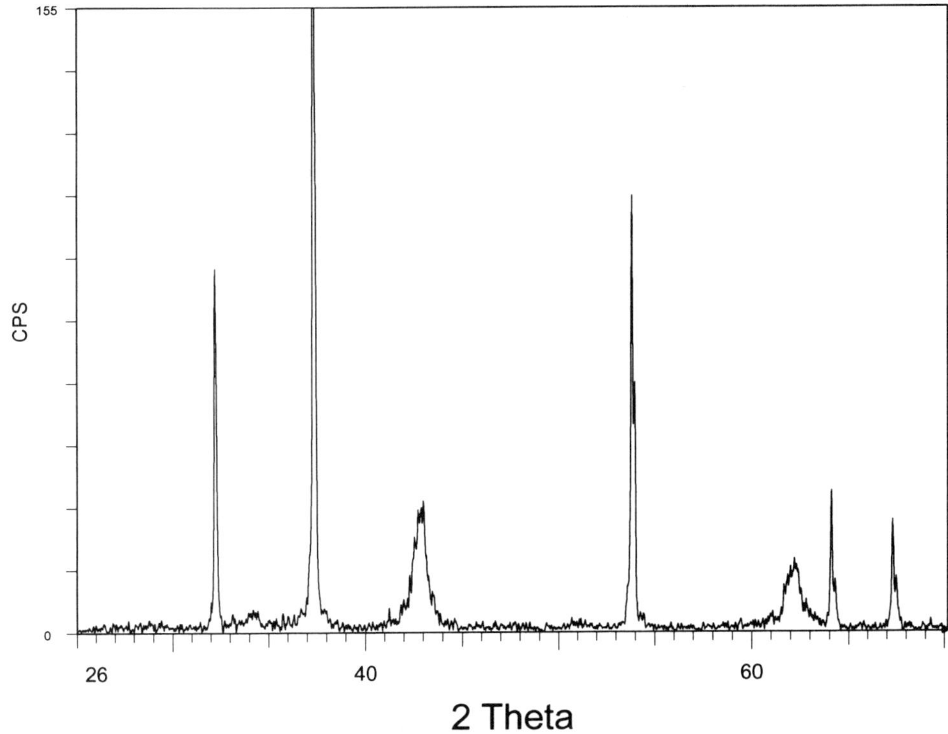

FIG. 7.2. X-ray diffraction pattern of an MgO–CaO mixture.

it is better to avoid these reflections to estimate the phase proportions. Calculate the R values, using Eq. (7.8) for the MgO and CaO phases, and tabulate your results in Tables 7.3 and 7.4. Remember to calculate the required values from the calculated lattice parameters and the structure factor, multiplicity, and Lorentz–polarization values. Use the appropriate θ values for the reflections to calculate these factors. Ignore the temperature factor in your calculations.

RESULTS

Tabulate your results in Tables 7.3 and 7.4.

Integrated intensity of the $(200)_{\mathrm{MgO}}$ peak, I_{MgO} = 108.2
R value for $(200)_{\mathrm{MgO}}$ peak, R_{MgO} =
Integrated intensity of the $(220)_{\mathrm{MgO}}$ peak, I_{MgO} = 65.4
R value for $(220)_{\mathrm{MgO}}$ peak, R_{MgO} =
Integrated intensity of the $(111)_{\mathrm{CaO}}$ peak, I_{CaO} = 65.0
R value for $(111)_{\mathrm{CaO}}$ peak, R_{CaO} =
Integrated intensity of the $(220)_{\mathrm{CaO}}$ peak, $\mathrm{I}_{\mathrm{CaO}}$ = 93.2
R value for $(220)_{\mathrm{CaO}}$ peak, R_{CaO} =
Since

$$\frac{I_{\mathrm{MgO}}}{I_{\mathrm{CaO}}} = \frac{R_{\mathrm{MgO}}\, c_{\mathrm{MgO}}}{R_{\mathrm{CaO}}\, c_{\mathrm{CaO}}}, \quad \frac{c_{\mathrm{MgO}}}{c_{\mathrm{CaO}}} = \frac{I_{\mathrm{MgO}}}{I_{\mathrm{CaO}}} \times \frac{R_{\mathrm{CaO}}}{R_{\mathrm{MgO}}}$$

and

$$c_{\mathrm{MgO}} + c_{\mathrm{CaO}} = 1$$

the values of c_{MgO} and c_{CaO} are

$$c_{\mathrm{MgO}} = \qquad\qquad \text{and} \qquad c_{\mathrm{CaO}} =$$

Calculate the volume fractions for both pairs of reflections and determine the average value.

EXERCISES

7.1. Assume that the intensity of a reflection in the powder method is directly proportional to F^2 and the amount of material present and that it is inversely proportional to μ, the linear absorption coefficient. Consider a mechanical mixture of two elements Cu ($Z = 29$ and $\mu = 46\ \mathrm{cm}^{-1}$ for Cu Kα radiation) and Au ($Z = 79$ and $\mu = 401\ \mathrm{cm}^{-1}$ for Cu Kα radiation) and

TABLE 7.3. Work Table for Calculation of R_{MgO} in an MgO–CaO Phase Mixture

Reflection	200	220
2θ (°)	42.91	62.30
Lattice parameter (nm), a	0.4213	0.4213
Volume of unit cell (nm^3), v		
$\dfrac{\sin\theta}{\lambda}$		
Atomic scattering factor, f (from Appendix 3)		
F^2 (use the appropriate equations from Appendix 4)		
Multiplicity factor, p (from Appendix 6)		
Lorentz–polarization factor $\left(\dfrac{1+\cos^2 2\theta}{\sin^2\theta\cos\theta}\right)$ (from Appendix 7)		
$R = \dfrac{1}{v^2}\left[F^2 p\left(\dfrac{1+\cos^2 2\theta}{\sin^2\theta\cos\theta}\right)\right]e^{-2M}$ [a]		

[a]We are ignoring the temperature factor e^{-2M} since the diffraction pattern is recorded at room temperature, we are using low-angle reflections, and because we are taking the ratios of R values.

TABLE 7.4. Work Table for Calculation of R_{CaO} in an MgO–CaO Phase Mixture

Reflection	111	220
2θ (°)	32.17	53.83
Lattice parameter (nm), a	0.4811	0.4811
Volume of unit cell (nm^3), v		
$\dfrac{\sin\theta}{\lambda}$		
Atomic scattering factor, f (from Appendix 3)		
F^2 (use the appropriate equations from Appendix 4)		
Multiplicity factor, p (from Appendix 6)		
Lorentz–polarization factor $\left(\dfrac{1+\cos^2 2\theta}{\sin^2\theta\cos\theta}\right)$ (from Appendix 7)		
$R = \dfrac{1}{v^2}\left[F^2 p\left(\dfrac{1+\cos^2 2\theta}{\sin^2\theta\cos\theta}\right)\right]e^{-2M}$ [a]		

[a]We are ignoring the temperature factor e^{-2M} since the diffraction pattern is recorded at room temperature, we are using low-angle reflections, and because we are taking the ratios of R values.

calculate the smallest weight percent of copper that can be detected in such a mixture, assuming that the intensity ratio must exceed 5% to be detectable above background.

7.2. Consider the intensity ratio of two reflections of components A and B in a binary mixture. Prepare a schematic plot of this ratio against the amount of component B present when $\mu_A > \mu_B$, when $\mu_A = \mu_B$, and when $\mu_A < \mu_B$.

7.3. Microscopic examination of hardened 1.0% carbon steel shows no undissolved carbides. X-ray examination of this steel in a diffractometer with Co $K\alpha$ radiation shows that the integrated intensity of the 311 austenite peak is 2.33 and the integrated intensity of the unresolved 112-211 martensite doublet is 16.32, both in arbitrary units. Calculate the volume percent of austenite in the steel. Use the lattice parameter values a (austenite) = 0.3599 nm, a (martensite) = 0.2854 nm, and c (martensite) = 0.2983 nm. Ignore the temperature factor.

EXPERIMENTAL MODULE 8

Identification of an Unknown Specimen

OBJECTIVE OF THE EXPERIMENT

To identify an unknown specimen by x-ray diffraction procedures.

MATERIALS REQUIRED

Any single-phase and any two-phase material.

BACKGROUND AND THEORY

X-ray diffraction is a convenient method to identify an unknown specimen by determining its crystal structure and comparing it with a repository of standard powder diffraction patterns. Exact matching of the interplanar spacings (*d* values) and intensities of all peaks between the standard and observed diffraction patterns confirms the identity of the unknown as the same as the standard or reference material. This was realized as early as 1919 by Hull (A. W. Hull, A New Method of Chemical Analysis, *J. Am. Chem. Soc.* **41** (1919), 1168) who wrote

> ...every crystalline substance gives a pattern; that the same substance always gives the same pattern; and that in a mixture of substances each produces its pattern independently of the others, so that the photograph obtained with a mixture is the superimposed sum of the photographs that would be obtained by exposing each of the components separately for the same length of time. This law applies quantitatively to the intensities of the

> lines, as well as to their positions, so that the method is capable of development as a quantitative analysis.

The simplicity and advantages of the powder diffraction method for identification of an unknown substance are as follows:

- The powder diffraction pattern is determined by the exact atomic arrangement in a material, so it is like a "fingerprint" of the material.
- Each substance in a mixture produces its own characteristic diffraction pattern independently of the others.
- The x-ray diffraction pattern discloses the presence of a substance as that substance actually exists in the specimen.
- Only a small amount of the material is required for recording the x-ray diffraction pattern.
- The method is nondestructive.

Since the x-ray diffraction pattern is like a fingerprint of any substance, it should be possible to unambiguously identify that material from its x-ray diffraction pattern. If we have a large collection of x-ray diffraction patterns from a number of substances, then the unknown material can be identified by obtaining its diffraction pattern and then determining which pattern from the collection matches exactly with the recorded x-ray diffraction pattern. The Powder Diffraction File (PDF) organized by the Joint Committee on Powder Diffraction Standards (JCPDS), later renamed the International Centre for Diffraction Data, has a present collection of nearly 80,000 standard diffraction patterns. The special features of these files (often called "cards") and the type of information obtainable from each card was described in Sec. 3.4 of Part I. Let's now see how we can identify an unknown specimen from these standard patterns.

X-ray diffraction patterns indicate the state in which a substance exists, e.g., as a pure element, or a compound, or a mechanical mixture, whereas conventional chemical analysis techniques reveal the presence of only the chemical constituents. For example, in a plain carbon steel specimen, conventional chemical analysis indicates the amount of carbon and iron in the specimen; but x-ray diffraction patterns indicate whether the phases present in the steel are ferrite, cementite, austenite, martensite, etc., or a combination of them. Furthermore, polymorphic modifications of substances can be conveniently identified by x-ray diffraction techniques since each polymorph produces its own characteristic diffraction pattern. Conventional chemical analysis, on the other hand, again only tells us about the chemical element(s) present, which is (are) the same in both polymorphs.

Even though we had earlier recommended that you use nanometers (nm) as the unit of distance (for interplanar spacings and lattice parameters), the PDF (still) lists all the interplanar spacings in units of ångstroms. Therefore, in this module we suggest that you use ångstroms for measuring the interplanar spacings.

The cards list the interplanar spacings (*d* values)—and not 2θ values since these depend on the wavelength of the x-rays used, while *d* is independent of the radiation used—(through the use of a computer you can quickly convert the *d* values to 2θ values for any given x-ray wavelength), Miller indices (*hkl*) of the planes having these spacings, and the relative intensities (I/I_1) of all the reflections observed, in addition to other crystallographic data (again see Sec. 3.4 in Part I for more details).

The International Centre for Diffraction Data publishes two indexes—alphabetical and numerical—and they are published separately for inorganic and organic phases. The alphabetical index manual lists all the compounds in alphabetical order, so if you have an idea or intelligent guess of what the material might be, you can look up the corresponding "card," compare the *d* spacings and relative intensities and confirm your guess. The numerical index manual (also called the Hanawalt Search Manual) lists the compounds in decreasing order of the *d* spacings of the eight most intense reflections. These *d* spacings are arranged in groups starting from 999.9 to 8.00 Å as one group, 7.99 to 7.00 Å as another group, etc., down to 1.37 to 1.00 Å. The complete grouping is listed in Table 10 in Sec. 3.4, Part I. The ± error range in each group is different and varies from –0.20 Å for the largest *d* spacing group (999.9 to 8.00 Å) to 0.01 Å for the smallest group (1.37 to 1.00 Å). The *d* value of the strongest peak in the pattern determines the group into which the entry falls, and the *d* value of the second strongest peak determines the subgroup. Since the accuracy in the measurement of *d* spacings (especially at low angles) is not very high, you should always look for a range of *d* spacings. For example, if the observed *d* spacing is 2.26 Å, then you should look for *d* values from 2.23 to 2.29 Å. You will find the numerical index manual or Hanawalt Search Manual the most useful in the identification of an unknown specimen from its x-ray diffraction pattern.

Identification of an unknown specimen begins with recording its x-ray diffraction pattern with a suitable radiation and under suitable voltage and current conditions. You should ensure that the specimen, if in powder form, has the ideal fine grain size to give a good x-ray diffraction pattern (containing sharp and intense diffraction peaks with a minimum of background) and that there is no preferred orientation in the specimen.

It is desirable that you record the x-ray diffraction pattern with Cu Kα radiation since all the standard patterns in the PDF are recorded with this radiation, and it would be better to have the unknown and standard patterns recorded under similar conditions. (However, with modern computer software you can easily convert the data from one radiation to another. But you should realize that the relative intensities of the reflections can be different when different radiations are used. Further, you can have the card display either the 2θ values or the *d* spacings; *d* spacings are preferred because the search manuals list only *d* spacings.) Convert all 2θ values into interplanar spacings (*d* values) and list the relative integrated intensities of all the reflections (remember to subtract the background intensity), assuming that the most intense reflection has an intensity of 100%. Compute the *d* spacings to the nearest 0.01 Å for low-angle peaks and to 0.001 Å for high-angle peaks.

Identifying the unknown material involves the following steps:

1. Identify the three most intense reflections in the recorded pattern. Designate the interplanar spacing corresponding to the most intense peak by d_1, that for the next most intense by d_2, and the next one by d_3.
2. Locate the proper d_1 group in the Hanawalt Search Manual.
3. Once you have found a reasonable match for d_1, look for the closest match to d_2.
4. Repeat the procedure for d_3.
5. After the closest match has been found for d_1, d_2, and d_3, compare their relative integrated intensities with the tabulated values.
6. When the *d* values and the intensities for the three most intense reflections from the observed pattern match well with any of those listed in the manual, locate the proper card and compare the *d* and intensity values for *all* the reflections. When full agreement is obtained, identification is complete.

In practice, the unknown material may contain one or more phases. The procedure for identification, however, is the same in both cases except that if the material contains only one phase the identification is easy and relatively straightforward. Let's now see how an unknown material can be identified by using the x-ray diffraction technique and the Hanawalt Search Manual.

Single-Phase Material

The experimental 2θ, *d*, and I/I_1 (where *I* is the intensity of any peak and I/I_1 is the ratio of any given peak to that of the most intense peak, taken as 100%) values from the x-ray diffraction pattern (Fig. 8.1) of the unknown specimen recorded with Cu Kα radiation are listed in Table 8.1.

Note that the three most intense peaks in the pattern have *d* spacings 2.16, 2.50, and 1.53 Å with intensities 100, 85, and 55, respectively. In the Hanawalt Search Manual, you will find that the *d* spacing corresponding to the most intense peak is 2.22 to 2.16 (± 0.01) Å. The *d* spacing corresponding to the second most intense peak (2.50 Å) has many entries. Table 8.2 reproduces some of the entries, with the most intense peak having a *d* spacing of 2.16 Å and the second most intense peak 2.50 Å. In the Hanawalt Search Manual the *d* values of the three most intense peaks are shown in boldface.

In Table 8.2 notice that only three phases—$NaFe_2O_3$, TaN, and TiC—have *d* spacing of 1.53 Å for the third strongest reflection. Therefore, we now have to decide which of these three substances is our unknown specimen. To do this, compare the list of *d* spacings obtained from the x-ray diffraction pattern of the unknown specimen and intensities of the other reflections with those of the possible materials in Table 8.2. The unknown pattern has a *d* spacing of 1.30 Å for the fourth most intense reflection, so the specimen cannot be $NaFe_2O_3$, which has a *d* spacing of 1.56 Å for the fourth most intense reflection.

If we continue this procedure, the fifth most intense reflection has a *d* spacing of 0.97 Å for the unknown specimen. Comparing TaN and TiC, we see that only TiC has a *d* spacing of 0.97 Å and an intensity of 30% of the most intense reflection. This matches very well with the values for the unknown specimen. A comparison of the *d* values and intensities in the unknown pattern and those of TiC (PDF #32-1383) confirms that the unknown specimen is TiC.

TABLE 8.1. Experimental Data for Identifying an Unknown Specimen

Peak #	2θ (°)	*d* (Å)	I/I_1
1	35.84	2.503	85
2	41.73	2.163	100
3	60. 47	1.530	55
4	72.37	1.305	32
5	76.02	1.251	18
6	90.75	1.082	11
7	101.71	0.993	13
8	105.48	0.968	25
9	121.27	0.884	25
10	135.17	0.833	15

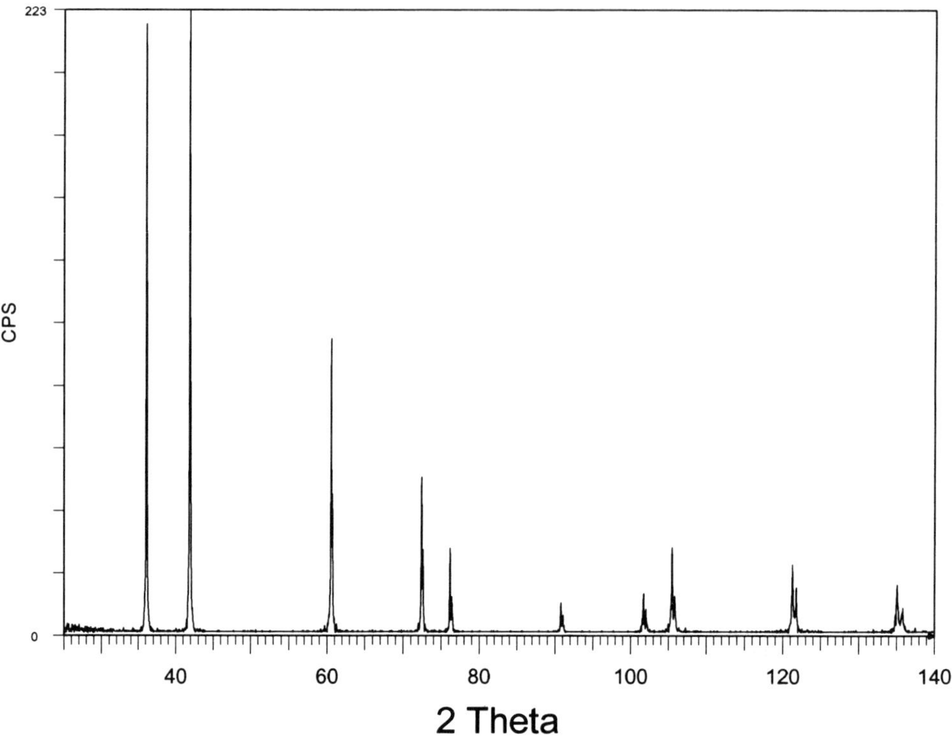

FIG. 8.1. X-ray diffraction pattern of the unknown single-phase material.

The preceding description reminds us that in order to unambiguously identify the material it is essential that *all* peaks in the experimental diffraction pattern must match the peaks listed in the standard diffraction pattern. In this situation the intensities also match very well, although sometimes they might be slightly different in the two patterns.

TABLE 8.2. Excerpt from the Hanawalt Search Manual (1996 edition, p. 1092)

Strongest reflections								PSC	Chemical formula	PDF#
$\mathbf{2.18_x}$	$\mathbf{2.50_5}$	$\mathbf{2.49_5}$	2.27_5	2.19_5	1.68_5	1.44_5	1.30_5	oP12.50	$C_{20}Cr_{25}V_{55}$	18-401
$\mathbf{2.18_x}$	$\mathbf{2.50_4}$	$\mathbf{1.53_4}$	1.56_3	2.60_2	2.59_2	1.31_2	1.30_2	hR8	$NaFe_2O_3$	32-1066
$\mathbf{2.17_x}$	$\mathbf{2.50_7}$	$\mathbf{2.44_6}$	3.30_3	3.12_3	1.56_3	2.63_2	3.81_1	hP*	$Ga_3Ta_5O_x$	33-750
$\mathbf{2.17_8}$	$\mathbf{2.50_x}$	$\mathbf{1.53_4}$	1.31_4	1.25_3	0.00_1	0.00_1	0.00_1	cF8	TaN	32-1283
$\mathbf{2.16_8}$	$\mathbf{2.50_x}$	$\mathbf{3.40_8}$	2.33_8	2.11_8	1.74_8	1.53_8	2.74_5	oP20	$LaCrOS_2$	31-659
$\mathbf{2.16_8}$	$\mathbf{2.50_x}$	$\mathbf{2.19_x}$	1.38_5	1.25_5	1.93_1	2.32_1	2.27_1		$(Al_{25}Fe_{25}Nb_{50})$	23-1008
$\mathbf{2.16_x}$	$\mathbf{2.50_8}$	$\mathbf{1.53_6}$	1.30_3	0.97_3	0.88_3	1.25_2	0.83_2	cF8	TiC	32-1383
$\mathbf{2.15_x}$	$\mathbf{2.50_5}$	$\mathbf{2.28_5}$	1.32_x	1.23_7	1.29_6	1.89_5	1.36_5	cF116	$(Re_{0.65}Fe_{0.35})_{23}B_6$	19-619
$\mathbf{2.15_8}$	$\mathbf{2.50_x}$	$\mathbf{2.20_x}$	1.45_5	1.38_5	1.25_5	2.03_2	1.93_2		μ'-$Al_{23}Co_{23}Ta_{54}$	23-1007

An entry of 0.00_1 indicates that either there were no additional reflections present in the pattern or they were not indexed. The subscript 1 is included only for computer purposes, it has no physical meaning.

Two-Phase Material

Identification of phases in a two-phase mixture is a little more complex, but follows the same principles already outlined. An x-ray diffraction pattern is recorded from the material, and the d (in units of ångstroms) and I/I_1 values are listed for all reflections in the pattern. The d_1, d_2, and d_3 values of the three most intense reflections are noted and compared with the Hanawalt Search Manual and one of the phases is identified. A similar procedure is followed for identifying the remaining phase. This procedure can also be extended to materials with more than two phases, but it is much more complex and time consuming. It may be helpful to use the "Hanawalt work forms" on which a record can be kept of the sequence of steps taken in the process of the solution, thereby avoiding omissions and repetitions. Use of such a form is illustrated in the PDF Search Manual–Hanawalt Method published by the JCPDS–International Centre for Diffraction Data. Alternatively, you can use your own worksheet to record all the possibilities you have worked out and found to be unsatisfactory.

Let's now work out an example on identifying the constituent phases in a two-phase mixture. Table 8.3 lists the 2θ (though not required for the analysis), d, and I/I_1 values of all observed reflections from the diffraction pattern of the two-phase material recorded with Cu Kα radiation shown in Fig. 8.2.

TABLE 8.3. Diffraction Data from an Unknown Two-Phase Material

Peak #	2θ (°)	d (Å)	I/I_1
1	37.26	2.411	68
2	43.29	2.088	100
3	44.51	2.034	59
4	51.87	1.761	28
5	62.89	1.477	57
6	75.40	1.260	13
7	76.43	1.245	13
8	79.37	1.206	8
9	92.99	1.062	15
10	95.05	1.044	4
11	98.49	1.017	5
12	106.97	0.958	5
13	111.11	0.934	16
14	129.18	0.853	15

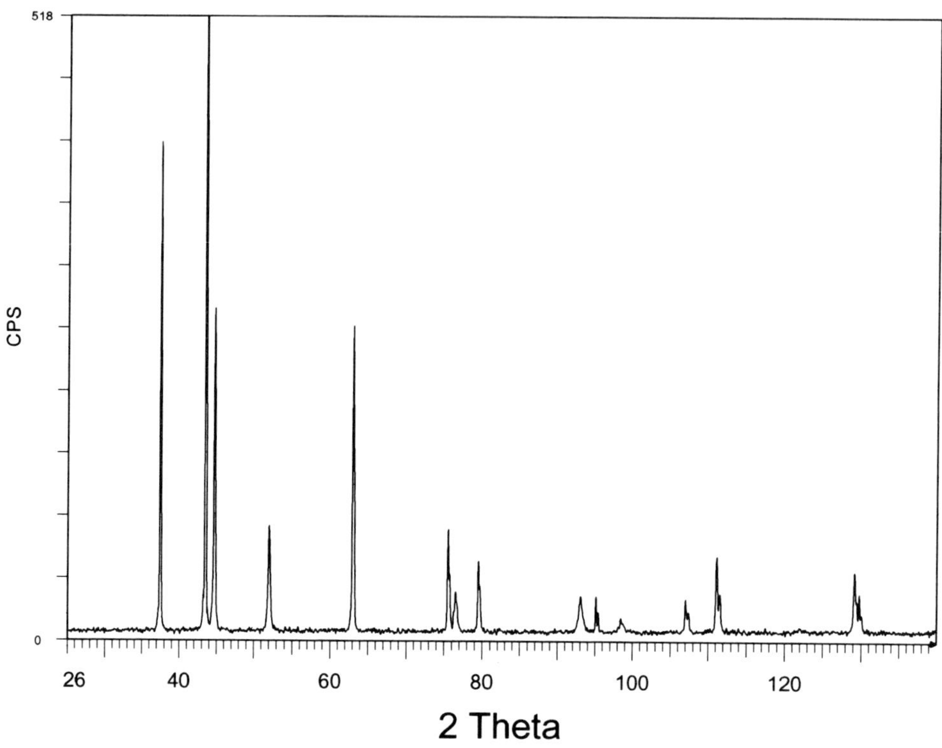

FIG. 8.2. X-ray diffraction pattern of the unknown two-phase material.

From Table 8.3 we note that the three strongest peaks in the pattern have *d* spacings 2.09, 2.41, and 2.03 Å. In the Hanawalt Search Manual, we find that 2.09 Å lies between of 2.15 and 2.09 (±0.1) Å. The excerpt from the Hanawalt Search Manual in Table 8.4 gives 10 possible materials, with the first two most intense spacings given as 2.09 and 2.41 Å. However, none of these have a *d* spacing of 2.03 Å for the third most intense reflection, which is observed in the unknown pattern. Since the present diffraction pattern is obtained from a mixture of two phases, it is possible that the third most intense peak observed in our diffraction pattern is the most intense peak from the second phase and does not belong to the first phase. Assuming that this is the case, we need to decide which of these 10 substances forms the first phase.

From Table 8.4 notice that four substances have *d* spacings of 2.09 Å and 2.41 Å for the first two most intense reflections, namely Rh_2Si, $TmAg_3$, NiO, and NiO (Bunsenite). Considering Rh_2Si as a possibility, we note that the third most intense peak has a *d* value of 2.23 Å and a relative intensity

TABLE 8.4. Excerpt from the Hanawalt Search Manual (1996 edition, p. 1145)

Strongest reflections								PSC	Chemical formula	PDF#
$\mathbf{2.12}_x$	$\mathbf{2.41}_x$	$\mathbf{3.87}_3$	1.43_2	1.35_2	0.00_1	0.00_1	0.00_1		$Cu_{0.139}Mn_{0.861}O_2$	41-183
$\mathbf{2.12}_9$	$\mathbf{2.41}_x$	$\mathbf{2.22}_x$	1.80_9	1.87_8	1.83_8	1.56_8	2.81_5	oI36	FeGeNb	24-535
$\mathbf{2.11}_x$	$\mathbf{2.41}_5$	$\mathbf{3.27}_4$	2.23_2	1.32_1	1.73_1	1.63_1	1.41_1	oP4	IrW	29-691
$\mathbf{2.09}_x$	$\mathbf{2.41}_x$	$\mathbf{2.23}_x$	2.13_x	2.92_8	2.24_8	2.18_8	1.97_8	oP12	Rh_2Si	37-1316
$\mathbf{2.09}_x$	$\mathbf{2.41}_x$	$\mathbf{1.48}_x$	1.27_x	0.96_8	0.86_8	0.81_8	1.21_6	cP4	$TmAg_3$	27-626
$\mathbf{2.09}_x$	$\mathbf{2.41}_6$	$\mathbf{1.48}_6$	1.26_1	1.21_1	0.93_1	0.85_1	1.04_1	hR2	NiO	44-1159
$\mathbf{2.09}_x$	$\mathbf{2.41}_9$	$\mathbf{1.48}_6$	0.93_2	0.85_2	1.26_2	1.21_1	1.04_1	cF8	NiO (Bunsenite)	4-835
$\mathbf{2.08}_x$	$\mathbf{2.41}_4$	$\mathbf{2.92}_3$	2.04_3	1.86_2	1.32_2	1.21_2	1.18_2	hP38	$Co_{17}Gd_2$	43-1383
$\mathbf{2.08}_x$	$\mathbf{2.41}_9$	$\mathbf{1.47}_6$	1.26_3	0.93_3	0.85_2	1.20_2	0.96_2	cP60	V_8C_7	35-786
$\mathbf{2.15}_x$	$\mathbf{2.40}_x$	$\mathbf{2.05}_x$	1.33_x	1.22_x	1.70_8	1.42_8	1.36_8	tI34	$Pt_{12}Si_5$	34-903

over 100%. In our diffraction pattern we do not have a *d* spacing of 2.23 Å. Therefore, the first phase is definitely not Rh_2Si. Considering the other possibilities, we note that the third peak has a *d* value of 1.48 Å in the three remaining cases. The relative intensity of the third peak for $TmAg_3$ is over 100%, whereas for NiO (hR2), and Bunsenite (synthetic NiO) (cF8) it is 60. Since the reflection with 1.48 Å in our pattern has an intensity of 57, it is possible that the phase is either NiO(hR2) or NiO(cF8) but not $TmAg_3$. The fourth most intense peak in NiO has a *d* value of 1.26 Å and an intensity of 10, whereas for Bunsenite these values are 0.93 Å and 20. When we compare our data with those for NiO and Bunsenite, we realize that complete matching is obtained only for Bunsenite. Now, let's look at the whole pattern of Bunsenite (card #4-835) and compare it with the observed pattern. The results are summarized in Table 8.5.

From Table 8.5 it is clear that all the expected *d* values of Bunsenite are present in the actual pattern. Further, the relative intensities of most peaks in both patterns also match. However, the intensity of the first peak with a *d* spacing of 2.41 Å has an intensity of 68 in the observed pattern, whereas it has 91 in the PDF. The most recent card (47-1049, yet to be published) for Bunsenite notes the intensity of this peak as 61. (Other differences exist between cards 4-835 and 47-1049.) Also, since the *hkl* values in Table 8.5 are unmixed, the material must be based on a face-centered cubic Bravais lattice. This additional factor also confirms that the phase is Bunsenite (cF8) and not the hexagonal polymorph of NiO (hR2). Thus, it can be concluded that the cubic form of NiO (Bunsenite) is one of the phases in the mixture. Identification of the unknown specimen will be complete when we assign all the remaining peaks to the other phase. These remaining peaks are listed in Table 8.6, with the

TABLE 8.5. Identification of the Unknown Two-Phase Material

Peak #	$d_{obs.}$ (Å)	$I/I_{1obs.}$	$d_{(Bunsenite)}$ (Å) (card #4-835)	$I/I_{1(Bunsenite)}$ (card #4-835)	$hkl_{(Bunsenite)}$
1	2.411	68	2.410	91	111
2	2.088	100	2.088	100	200
3	2.034	59			
4	1.761	28			
5	1.477	57	1.476	57	220
6	1.260	13	1.259	16	311
7	1.245	13			
8	1.206	8	1.206	13	222
9	1.062	15			
10	1.044	4	1.044	8	400
11	1.017	5			
12	0.958	5	0.958	7	331
13	0.934	16	0.934	21	420
14	0.853	15		17	422

intensities normalized so that the most intense peak of the remaining peaks has an intensity of 100%.

From the Hanawalt Search Manual for 2.03, 1.76, and 1.25 Å as the *d* values of the three strongest peaks, there are three possibilities—$FeNi_3$, Ni, and Ni_3Si (Table 8.7). Even though the first four *d* spacings of the unknown pattern match those of Fe Ni_3, Ni, and Ni_3Si in Table 8.7, the intensities are slightly different. The observed intensity for the second reflection in the unknown pattern is 47, and it is 60 for $FeNi_3$, 40 for Ni, and 70 for Ni_3Si. The intensities of the other reflections for $FeNi_3$ and N_3Si are consistently higher than those of our unknown. Therefore, we can consider Ni as the best choice. By comparison with card #4-850 (for nickel) the *d* values in the pattern can be indexed as arising from the (111), (200), (220), (311), and (222) planes, those typical of a face-centered cubic lattice. This fact lends additional support to our identification, since

TABLE 8.6. Data for Identification of the Second Phase

Peak #	*d* (Å)	I/I_1 observed	I/I_1 normalized
3	2.034	59	100
4	1.761	28	47
7	1.245	13	22
9	1.062	15	25
11	1.017	5	8

TABLE 8.7. Excerpt from the Hanawalt Search Manual (1996 edition, p. 1155)

Strongest reflections								PSC	Chemical formula	PDF#
2.04_x	**1.77_9**	**2.30_7**	3.00_3	1.27_3	3.44_1	2.50_1	2.17_1	h*30	$CaTiO_3$	40-43
2.04_x	**1.77_6**	**1.25_3**	1.07_4	1.02_1	0.81_1	0.79_1	0.00_1	cP4	$FeNi_3$	34-419
2.02_x	**1.77_6**	**1.75_4**	0.80_5	0.78_5	2.62_3	3.70_3	2.34_2	cF116	$B_6Li_3Ni_{20}$	42-967
2.07_9	**1.76_x**	**2.93_9**	3.39_9	1.19_7	0.82_7	0.81_7	1.69_6	cP12	$PdP_{0.67}S_{1.33}$	25-604
2.06_x	**1.76_8**	**2.24_4**	1.26_x	1.25_x	1.18_6	1.56_5	1.33_5		Mn_3Co_7	18-407
2.03_x	**1.76_4**	**1.25_2**	1.06_2	0.79_2	0.81_1	1.02_1	0.88_1	cF4	Ni	4-850
2.08_x	**1.75_x**	**3.57_8**	3.24_8	2.25_7	3.33_6	3.61_5	2.11_5	hP26	$As_4Cu_{3.2}Li_{5.8}$	23-801
2.06_x	**1.75_8**	**1.19_8**	1.33_7	1.12_7	0.98_7	0.92_7	3.37_6	tP7	$CdIn_2Se_4$	8-267
2.02_x	**1.75_7**	**1.23_7**	1.06_7	0.87_7	0.80_7	1.01_5	0.88_5	cP4	Ni_3Si	6-690
2.04_x	**1.74_x**	**3.33_8**	1.34_8	1.18_8	1.11_8	0.98_8	2.89_5	cF*	LaOF	17-280
2.04_x	**1.74_8**	**3.32_5**	2.88_4	1.66_1	1.44_1	0.00_1	0.00_1	cF10	Ni_3Te_2	23-1279

Ni has a face-centered cubic lattice, whereas $FeNi_3$ and Ni_3Si have primitive cubic lattices. The actual and expected *d* values and intensities match very well for Ni. Hence, it can be concluded that the second phase in the mixture is nickel. Thus, the starting material is a mixture of nickel and the cubic form of NiO (Bunsenite).

In identifying unknown specimens (whether single phase or multiphase) with this technique, remember that the *d* values and relative intensities of *all* the peaks must be accounted for. Further, all peaks expected of the substances must be present in the observed diffraction pattern.

In practice, you will not know the number of phases in an unknown specimen. In that case, you may first assume that the material contains only one phase and check whether you can match the observed *d* spacings and intensities of *all* the reflections with any of those in the standard collection. If matching is not obtained, then you may assume that the material contains more than one phase and proceed to identify the constituent phases.

Material identification by x-ray diffraction using the Hanawalt method has some limitations. First, in practice the intensities of all the peaks may not always match those listed on the standard cards. One reason is that the material may have developed texture (preferred orientation). In this case, the intensities are higher for some reflections and lower for some others, depending on the type of preferred orientation. Second, if the experimental diffraction pattern is not recorded in the full range of 2θ, and if some of the high-angle peaks are more intense than those at low angles, the sequence of *d* spacings with decreasing intensity may be wrong. Third, it is also possible that the cards in the PDF have some errors. One such error was found in the diffraction pattern of titanium (PDF

#5-682) where the 210 reflection had been omitted from the indexing. This error was corrected only in 1997 (PDF #44-1294). If errors exist in the cards, the identification may also be in error. Last, the method is totally inadequate for identification if the diffraction pattern of the substance is not included in the PDF!

EXPERIMENTAL PROCEDURE

An x-ray diffraction pattern of an unknown single-phase material recorded with Cu Kα radiation is presented in Fig. 8.3. The 2θ, *d* values (in Å), and relative intensities of all peaks in the diffraction pattern are listed in Table 8.8. Identify the phase in the specimen, using the procedures discussed here. Additionally, you may take any unlabeled powder from the shelf, record the x-ray diffraction pattern by Cu Kα radiation, and identify the phase(s) in the specimen using these procedures.

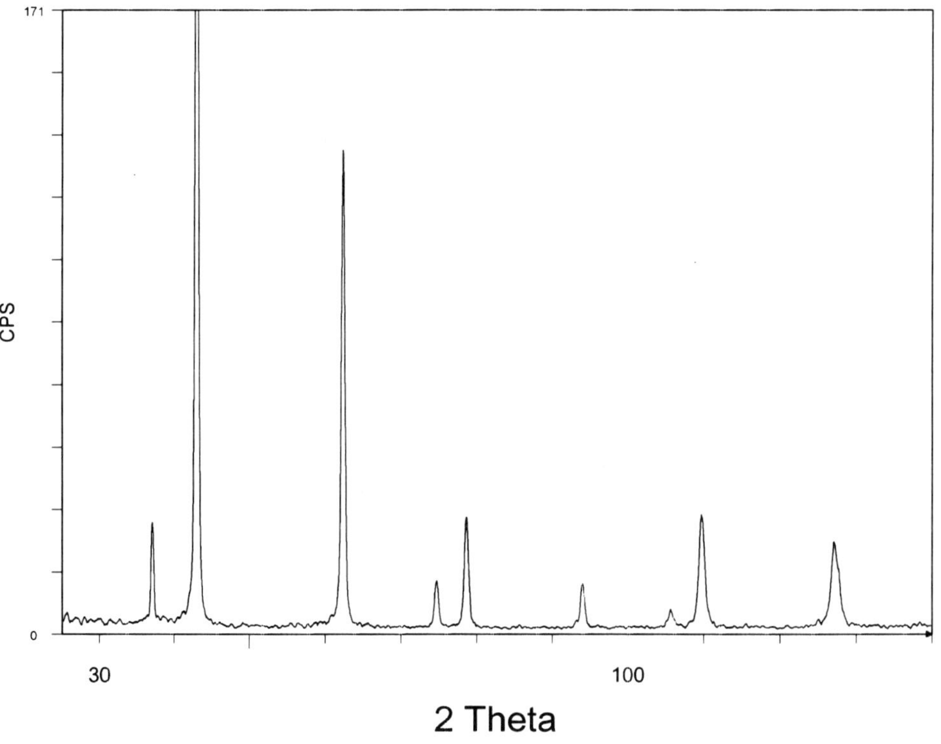

FIG. 8.3. X-ray diffraction pattern of an unknown material.

TABLE 8.8. Data for Identification of the Unknown Specimen

Peak #	2θ (°)	d (Å)	I/I_1
1	36.90	2.434	15
2	42.79	2.111	100
3	62.12	1.493	51
4	74.49	1.273	5
5	78.49	1.217	13
6	93.93	1.054	6
7	105.71	0.966	2
8	109.84	0.941	16
9	127.16	0.860	14

EXERCISES

8.1. An unknown single-phase substance has the following d and I/I_1 values. Identify the substance by reference to the PDF and the Hanawalt Search Manual.

d (Å)	I/I_1	d (Å)	I/I_1
3.123	100	1.209	2
2.705	10	1.103	10
1.912	51	1.040	5
1.633	30	0.9597	3
1.561	5	0.9138	5
1.351	6	0.8548	3
1.240	10	0.8244	2

8.2. An unknown single-phase substance has the following d and I/I_1 values. Identify the substance by reference to the PDF and the Hanawalt Search Manual.

d (Å)	I/I_1	d (Å)	I/I_1
3.155	92	1.222	1
2.731	1	1.115	17
1.932	100	1.051	7
1.647	33	0.966	4
1.577	1	0.911	1
1.366	10	0.864	8
1.253	9	0.833	3

Appendixes

APPENDIX 1: PLANE-SPACING EQUATIONS AND UNIT CELL VOLUMES

Plane Spacings

The distance, d, between adjacent planes in the set (hkl) may be found from the following equations where a, b, c, α, β, and γ represent the lattice parameters of the unit cell.

Cubic:

$$\frac{1}{d^2} = \frac{h^2 + k^2 + l^2}{a^2}$$

Rhombohedral:

$$\frac{1}{d^2} = \frac{(h^2 + k^2 + l^2)\sin^2\alpha + 2\,(hk + kl + hl)\,(\cos^2\alpha - \cos\alpha)}{a^2\,(1 - 3\cos^2\alpha + 2\cos^3\alpha)}$$

Hexagonal:

$$\frac{1}{d^2} = \frac{4}{3}\left(\frac{h^2 + hk + k^2}{a^2}\right) + \frac{l^2}{c^2}$$

Tetragonal:

$$\frac{1}{d^2} = \frac{h^2 + k^2}{a^2} + \frac{l^2}{c^2}$$

Orthorhombic:

$$\frac{1}{d^2} = \frac{h^2}{a^2} + \frac{k^2}{b^2} + \frac{l^2}{c^2}$$

Monoclinic:

$$\frac{1}{d^2} = \frac{1}{\sin^2 \beta}\left(\frac{h^2}{a^2} + \frac{k^2 \sin^2 \beta}{b^2} + \frac{l^2}{c^2} - \frac{2hl \cos \beta}{ac}\right)$$

Triclinic:

$$\frac{1}{d^2} = \frac{1}{V^2}(S_{11} h^2 + S_{22} k^2 + S_{33} l^2 + 2S_{12} hk + 2S_{23} kl + 2S_{13} hl)$$

In the equation for the triclinic lattice,

V = volume of the unit cell
$S_{11} = b^2c^2 \sin^2 \alpha$
$S_{22} = a^2c^2 \sin^2 \beta$
$S_{33} = a^2b^2 \sin^2 \gamma$
$S_{12} = abc^2(\cos \alpha \cos \beta - \cos \gamma)$
$S_{23} = a^2bc(\cos \beta \cos \gamma - \cos \alpha)$
$S_{13} = ab^2c(\cos \gamma \cos \alpha - \cos \beta)$

Unit-Cell Volumes

The volume of the unit cell may be determined by using the following equations:

Cubic:

$$V = a^3$$

Rhombohedral:

$$V = a^3 \sqrt{1 - 3\cos^2 \alpha + 2\cos^3 \alpha}$$

Hexagonal:

$$V = \frac{\sqrt{3}a^2c}{2} = 0.866a^2c$$

Tetragonal:

$$V = a^2c$$

Orthorhombic:

$$V = abc$$

Monoclinic:

$$V = abc \sin \beta$$

Triclinic:

$$V = abc\sqrt{1 - \cos^2 \alpha - \cos^2 \beta - \cos^2 \gamma + 2 \cos \alpha \cos \beta \cos \gamma}$$

APPENDIX 2: QUADRATIC FORMS OF MILLER INDICES FOR THE CUBIC SYSTEM

$h^2 + k^2 + l^2$	*hkl*	Lattice[a]
1	100	P
2	110	P, I
3	111	P, F, D
4	200	P, I, F
5	210	P
6	211	P, I
7		
8	220	P, I, F, D
9	300, 221	P
10	310	P, I
11	311	P, F, D
12	222	P, I, F
13	320	P
14	321	P, I
15		
16	400	P, I, F, D
17	410, 322	P
18	411, 330	P, I
19	331	P, F, D
20	420	P, I, F
21	421	P
22	332	P, I
23		
24	422	P, I, F, D
25	500, 430	P
26	510, 431	P, I
27	511, 333	P, F, D
28		
29	520, 432	P
30	521	P, I
31		
32	440	P, I, F, D
33	522, 441	P
34	530, 433	P, I
35	531	P, F, D
36	600, 442	P, I, F
37	610	P
38	611, 532	P, I
39		
40	620	P, I, F, D

[a]P = primitive cubic, I = body-centered cubic, F = face-centered cubic, D = diamond cubic. Even though diamond cubic is not one of the Bravais lattices, we included it here separately for convenience only.

APPENDIX 3: ATOMIC AND IONIC SCATTERING FACTORS OF SOME SELECTED ELEMENTS

$(\sin\theta)/\lambda$ (nm^{-1})	0.0	1.0	2.0	3.0	4.0	5.0	6.0	7.0	8.0	9.0	10.0
H	1	0.811	0.481	0.251	0.130	0.071	0.040	0.024	0.015	0.010	0.007
Li	3	2.215	1.742	1.513	1.270	1.033	0.823	0.650	0.512	0.404	0.320
C	6	5.107	3.560	2.494	1.948	1.685	1.537	1.426	1.322	1.219	1.114
N	7	6.180	4.563	3.219	2.393	1.942	1.697	1.551	1.445	1.353	1.265
O	8	7.245	5.623	4.089	3.006	2.338	1.946	1.714	1.568	1.463	1.377
Na^{+}	10	9.546	8.374	6.894	5.471	4.290	3.395	2.753	2.305	1.997	1.785
Na	11	9.760	8.335	6.881	5.471	4.293	3.398	2.754	2.305	1.997	1.784
Mg^{+2}	10	9.662	8.735	7.513	6.210	5.025	4.046	3.288	2.724	2.315	2.023
Mg	12	10.47	8.75	7.446	6.194	5.034	4.059	3.297	2.729	2.317	2.022
Al^{+3}	10	9.738	9.011	7.975	6.813	5.683	4.681	3.851	3.195	2.693	2.319
Al	13	11.23	9.158	7.873	6.766	5.692	4.713	3.883	3.221	2.712	2.330
Si^{+4}	10	9.790	9.199	8.327	7.306	6.259	5.277	4.418	3.701	3.124	2.673
Si	14	12.13	9.673	8.231	7.202	6.240	5.312	4.470	3.750	3.164	2.702
Cl	17	15.23	11.99	9.576	8.181	7.305	6.595	5.915	5.245	4.607	4.023
Cl^{-}	18	15.69	11.99	9.524	8.162	7.305	6.600	5.920	5.248	4.608	4.024
Ca	20	17.33	14.30	11.71	9.650	8.275	7.392	6.762	6.228	5.717	5.209
Ti	22	19.41	16.04	13.20	10.85	9.148	8.007	7.240	6.676	6.200	5.752
Cr	24	21.79	18.26	14.97	12.23	10.18	8.756	7.791	7.118	6.606	6.172
Fe	26	23.68	20.05	16.74	13.85	11.50	9.753	8.512	7.645	7.023	6.545
Ni	28	25.81	22.15	18.70	15.58	12.96	10.91	9.392	8.301	7.519	6.944
Cu	29	27.08	23.54	19.87	16.51	13.71	11.51	9.861	8.663	7.799	7.166
Zn	30	27.93	24.28	20.72	17.42	14.56	12.24	10.44	9.108	8.132	7.417
Ga	31	28.68	24.94	21.48	18.28	15.41	13.00	11.07	9.604	8.510	7.702
Ge	32	29.53	25.57	22.14	19.05	16.23	13.77	11.75	10.15	8.937	8.028
As	33	30.47	26.24	22.72	19.73	16.99	14.54	12.44	10.74	9.411	8.396
Ag	47	43.96	38.15	32.42	27.71	24.18	21.61	19.66	18.07	16.65	15.32
Cd	48	44.80	38.93	33.25	28.47	24.78	22.06	20.03	18.41	17.00	15.70
Te	52	48.17	41.62	36.08	31.42	27.46	24.23	21.71	19.78	18.26	16.99
Cs	55	50.63	43.89	37.90	33.24	29.38	26.07	23.30	21.07	19.31	17.90
W	74	69.78	62.52	55.54	49.21	43.69	38.96	34.88	31.33	28.24	25.58
Au	79	75.14	67.30	59.40	52.58	46.93	42.21	38.16	34.58	31.39	28.53
Po	84	79.31	70.87	63.09	56.28	50.37	45.34	41.09	37.43	34.22	31.34

From *International Tables for Crystallography*, Volume C, edited by A. J. C. Wilson, Kluwer Academic Publishers, Dordrecht, The Netherlands, 1995, pp. 477–485.

APPENDIX 4: SUMMARY OF STRUCTURE FACTOR CALCULATIONS

Pearson Symbol and Typical Examples	Atom Coordinates	F^2
cP1 (α-Po)	0,0,0	$F^2 = f^2$ for all *hkl* values
cP2 (CsCl, β-CuZn)	Cs: 0,0,0; Cl: $\frac{1}{2},\frac{1}{2},\frac{1}{2}$	$F^2 = (f_{Cs} + f_{Cl})^2$ for $h + k + l$ even
		$F^2 = (f_{Cs} - f_{Cl})^2$ for $h + k + l$ odd
cP4 (Cu_3Au)	Au: 0,0,0	$F^2 = (f_{Au} + 3f_{Cu})^2$ for *hkl* unmixed
	Cu: $\frac{1}{2},\frac{1}{2},0$; $\frac{1}{2},0,\frac{1}{2}$; $0,\frac{1}{2},\frac{1}{2}$	$F^2 = (f_{Au} - f_{Cu})^2$ for *hkl* mixed
cI2 (W, Cr, α-Fe)	0,0,0; $\frac{1}{2},\frac{1}{2},\frac{1}{2}$	$F^2 = 4f^2$ for $h + k + l$ even
		$F^2 = 0$ for $h + k + l$ odd
cF4 (Al, Cu, Ni)	0,0,0; $\frac{1}{2},\frac{1}{2},0$; $\frac{1}{2},0,\frac{1}{2}$; $0,\frac{1}{2},\frac{1}{2}$	$F^2 = 16f^2$ for *hkl* unmixed
		$F^2 = 0$ for *hkl* mixed
cF8 (Si, Ge)	0,0,0; $\frac{1}{2},\frac{1}{2},0$; $\frac{1}{2},0,\frac{1}{2}$; $0,\frac{1}{2},\frac{1}{2}$	$F^2 = 64f^2$ for $h + k + l = 4N$
	$\frac{1}{4},\frac{1}{4},\frac{1}{4}$; $\frac{3}{4},\frac{3}{4},\frac{1}{4}$; $\frac{3}{4},\frac{1}{4},\frac{3}{4}$; $\frac{1}{4},\frac{3}{4},\frac{3}{4}$	$F^2 = 32f^2$ for $h + k + l$ odd
		$F^2 = 0$ for *hkl* mixed
		$F^2 = 0$ for $h + k + l = 4N \pm 2$
cF8 (NaCl, TiN)	Na: 0,0,0; $\frac{1}{2},\frac{1}{2},0$; $\frac{1}{2},0,\frac{1}{2}$; $0,\frac{1}{2},\frac{1}{2}$	$F^2 = 16(f_{Na} + f_{Cl})^2$ for *hkl* even
	Cl: $\frac{1}{2},0,0$; $0,\frac{1}{2},0$; $0,0,\frac{1}{2}$; $\frac{1}{2},\frac{1}{2},\frac{1}{2}$	$F^2 = 16(f_{Na} - f_{Cl})^2$ for *hkl* odd
		$F^2 = 0$ for *hkl* mixed
cF8 (ZnS, GaAs)	S: 0,0,0; $\frac{1}{2},\frac{1}{2},0$; $\frac{1}{2},0,\frac{1}{2}$; $0,\frac{1}{2},\frac{1}{2}$	$F^2 = 16(f_s + f_{Zn})^2$ for $h + k + l = 4N$
	Zn: $\frac{1}{4},\frac{1}{4},\frac{1}{4}$; $\frac{3}{4},\frac{3}{4},\frac{1}{4}$; $\frac{3}{4},\frac{1}{4},\frac{3}{4}$; $\frac{1}{4},\frac{3}{4},\frac{3}{4}$	$F^2 = 16(f_s - f_{Zn})^2$ for $h + k + l = 2N$
		$F^2 = 16(f_s^2 + f_{Zn}^2)$ for $h + k + l$ odd
		$F^2 = 0$ for *hkl* mixed
hP2 (Zn, Mg, Ti)	0,0,0; $\frac{2}{3},\frac{1}{3},\frac{1}{2}$	$F^2 = 4f^2$ for $h + 2k = 3N$ and l even
		$F^2 = 3f^2$ for $h + 2k = 3N \pm 1$ and l odd
		$F^2 = f^2$ for $h + 2k = 3N \pm 1$ and l even
		$F^2 = 0$ for $h + 2k = 3N$ and l odd

APPENDIX 5: MASS ABSORPTION COEFFICIENTS μ/ρ (cm²/g) AND DENSITIES ρ (g/cm³) OF SOME SELECTED ELEMENTS

Absorber	Atomic Number	Density (g/cm³)	μ/ρ Cu Kα λ = 0.154184 nm	μ/ρ Cu Kβ λ = 0.139222 nm
H	1	0.08375×10^{-3}	0.391	0.388
Li	3	0.533	0.500	0.412
C	6	2.27	4.51	3.33
N	7	1.165×10^{-3}	7.44	5.48
O	8	1.332×10^{-3}	11.5	8.42
Na	11	0.966	29.7	22.0
Mg	12	1.74	40.0	29.6
Al	13	2.70	49.6	36.8
Si	14	2.33	63.7	47.5
Cl	17	3.214×10^{-3}	106	79.5
Ca	20	1.53	170	129
Ti	22	4.51	200	152
Cr	24	7.19	247	185
Fe	26	7.87	302	232
Ni	28	8.91	48.8	279
Cu	29	8.93	51.8	39.2
Zn	30	7.13	57.9	43.8
Ga	31	5.91	62.1	47.0
Ge	32	5.32	67.9	51.4
As	33	5.78	74.7	56.5
Ag	47	10.50	213	163
Cd	48	8.65	222	169
Te	52	6.25	267	204
Cs	55	1.91 (−10°C)	317	243
W	74	19.25	168	130
Au	79	19.28	201	155
Po	84		254	196

The mass absorption coefficients are from *International Tables for Crystallography*, Volume C, edited by A. J. C. Wilson, Kluwer Academic Publishers, Dordrecht, The Netherlands, 1995, pp. 200–205.

APPENDIX 6: MULTIPLICITY FACTORS

Cubic System

Miller Indices	*hkl*	*hhl*	0*kl*	0*kk*	*hhh*	00*l*
Example	345	112	012	011	111	002
Multiplicity	48	24	24	12	8	6

Hexagonal System

Miller Indices	*hkl*	*hhl*	0*kl*	*hk*0	*hh*0	0*k*0	00*l*
Example	123	112	012	120	110	010	002
Multiplicity	24	12	12	12	6	6	2

APPENDIX 7: LORENTZ–POLARIZATION FACTOR $\left(\frac{1+\cos^2 2\theta}{\sin^2\theta\cos\theta}\right)$

θ°	0.0	0.1	0.2	0.3	0.4	0.5	0.6	0.7	0.8	0.9
2	1639	1486	1354	1239	1138	1048	968.9	898.3	835.1	778.4
3	727.2	680.9	638.8	600.6	565.6	533.6	504.3	477.3	452.4	429.3
4	408.0	388.3	369.9	352.8	336.8	321.9	308.0	294.9	282.7	271.1
5	260.3	250.1	240.5	231.4	222.9	214.7	207.1	199.8	192.9	186.3
6	180.1	174.1	168.5	163.1	158.0	153.1	148.4	144.0	139.7	135.6
7	131.7	128.0	124.4	120.9	117.6	114.4	111.4	108.5	105.6	102.9
8	100.3	97.80	95.37	93.03	90.78	88.60	86.50	84.48	82.52	80.63
9	78.79	77.02	75.31	73.65	72.05	70.50	68.99	67.53	66.11	64.74
10	63.41	62.12	60.86	59.65	58.46	57.32	56.20	55.11	54.06	53.03
11	52.03	51.06	50.12	49.20	48.30	47.43	46.58	45.75	44.94	44.16
12	43.39	42.64	41.91	41.20	40.50	39.82	39.16	38.52	37.88	37.27
13	36.67	36.08	35.50	34.94	34.39	33.85	33.33	32.81	32.31	31.82
14	31.34	30.87	30.41	29.95	29.51	29.08	28.66	28.24	27.83	27.44
15	27.05	26.66	26.29	25.92	25.56	25.21	24.86	24.52	24.19	23.86
16	23.54	23.23	22.92	22.61	22.32	22.02	21.74	21.46	21.18	20.91
17	20.64	20.38	20.12	19.87	19.62	19.38	19.14	18.90	18.67	18.44
18	18.22	18.00	17.78	17.57	17.36	17.15	16.95	16.75	16.56	16.36
19	16.17	15.99	15.80	15.62	15.45	15.27	15.10	14.93	14.76	14.60
20	14.44	14.28	14.12	13.97	13.81	13.66	13.52	13.37	13.23	13.09
21	12.95	12.81	12.67	12.54	12.41	12.28	12.15	12.03	11.90	11.78
22	11.66	11.54	11.43	11.31	11.20	11.09	10.98	10.87	10.76	10.65
23	10.55	10.45	10.34	10.24	10.15	10.05	9.952	9.857	9.763	9.671
24	9.579	9.489	9.400	9.313	9.226	9.141	9.057	8.973	8.891	8.810
25	8.730	8.651	8.573	8.496	8.420	8.345	8.271	8.198	8.126	8.055
26	7.984	7.915	7.846	7.778	7.711	7.645	7.580	7.515	7.452	7.389
27	7.327	7.265	7.205	7.145	7.086	7.027	6.969	6.912	6.856	6.800
28	6.745	6.691	6.637	6.584	6.532	6.480	6.429	6.379	6.329	6.279
29	6.231	6.182	6.135	6.088	6.041	5.995	5.950	5.905	5.861	5.817
30	5.774	5.731	5.688	5.647	5.605	5.564	5.524	5.484	5.445	5.406
31	5.367	5.329	5.292	5.254	5.218	5.181	5.146	5.110	5.075	5.040
32	5.006	4.972	4.939	4.906	4.873	4.841	4.809	4.777	4.746	4.715
33	4.685	4.655	4.625	4.595	4.566	4.538	4.509	4.481	4.453	4.426
34	4.399	4.372	4.346	4.320	4.294	4.268	4.243	4.218	4.193	4.169
35	4.145	4.121	4.097	4.074	4.051	4.029	4.006	3.984	3.962	3.941
36	3.919	3.898	3.877	3.857	3.837	3.817	3.797	3.777	3.758	3.739
37	3.720	3.701	3.683	3.665	3.647	3.629	3.612	3.594	3.577	3.561
38	3.544	3.528	3.512	3.495	3.480	3.464	3.449	3.434	3.419	3.404
39	3.389	3.375	3.361	3.347	3.333	3.320	3.306	3.293	3.280	3.267
40	3.255	3.242	3.230	3.218	3.206	3.194	3.183	3.171	3.160	3.149
41	3.138	3.127	3.117	3.106	3.096	3.086	3.076	3.067	3.057	3.048
42	3.038	3.029	3.020	3.011	3.003	2.994	2.986	2.978	2.970	2.962
43	2.954	2.946	2.939	2.932	2.924	2.917	2.911	2.904	2.897	2.891
44	2.884	2.878	2.872	2.866	2.860	2.855	2.849	2.844	2.839	2.833
45	2.828	2.824	2.819	2.814	2.810	2.805	2.801	2.797	2.793	2.789

(continued)

Appendix 7. Continued

θ°	0.0	0.1	0.2	0.3	0.4	0.5	0.6	0.7	0.8	0.9
46	2.785	2.782	2.778	2.775	2.772	2.769	2.766	2.763	2.760	2.757
47	2.755	2.752	2.750	2.748	2.746	2.744	2.742	2.740	2.739	2.737
48	2.736	2.734	2.733	2.732	2.731	2.730	2.730	2.729	2.729	2.728
49	2.728	2.728	2.728	2.728	2.728	2.728	2.728	2.729	2.730	2.730
50	2.731	2.732	2.733	2.734	2.735	2.737	2.738	2.740	2.741	2.743
51	2.745	2.747	2.749	2.751	2.753	2.755	2.758	2.761	2.763	2.766
52	2.769	2.772	2.775	2.778	2.781	2.785	2.788	2.792	2.795	2.799
53	2.803	2.807	2.811	2.815	2.820	2.824	2.829	2.833	2.838	2.843
54	2.848	2.853	2.858	2.863	2.868	2.874	2.879	2.885	2.890	2.896
55	2.902	2.908	2.914	2.921	2.927	2.933	2.940	2.946	2.953	2.960
56	2.967	2.974	2.981	2.988	2.996	3.003	3.011	3.019	3.026	3.034
57	3.042	3.050	3.059	3.067	3.075	3.084	3.092	3.101	3.110	3.119
58	3.128	3.137	3.147	3.156	3.166	3.175	3.185	3.195	3.205	3.215
59	3.225	3.235	3.246	3.256	3.267	3.278	3.289	3.300	3.311	3.322
60	3.333	3.345	3.356	3.368	3.380	3.392	3.404	3.416	3.429	3.441
61	3.454	3.466	3.479	3.492	3.505	3.518	3.532	3.545	3.559	3.573
62	3.587	3.601	3.615	3.629	3.644	3.658	3.673	3.688	3.703	3.718
63	3.733	3.749	3.764	3.780	3.796	3.812	3.828	3.844	3.861	3.877
64	3.894	3.911	3.928	3.945	3.963	3.981	3.998	4.016	4.034	4.053
65	4.071	4.090	4.108	4.127	4.146	4.166	4.185	4.205	4.225	4.245
66	4.265	4.285	4.306	4.327	4.348	4.369	4.390	4.412	4.434	4.456
67	4.478	4.500	4.523	4.546	4.569	4.592	4.616	4.640	4.663	4.688
68	4.712	4.737	4.762	4.787	4.812	4.838	4.864	4.890	4.916	4.943
69	4.970	4.997	5.024	5.052	5.080	5.108	5.137	5.166	5.195	5.224
70	5.254	5.284	5.315	5.345	5.376	5.408	5.440	5.471	5.504	5.536
71	5.569	5.603	5.636	5.670	5.705	5.740	5.775	5.810	5.846	5.883
72	5.919	5.957	5.994	6.032	6.071	6.109	6.149	6.188	6.229	6.270
73	6.311	6.352	6.394	6.437	6.480	6.524	6.568	6.613	6.658	6.704
74	6.750	6.797	6.844	6.893	6.941	6.991	7.041	7.091	7.142	7.194
75	7.247	7.300	7.354	7.409	7.464	7.521	7.578	7.635	7.694	7.753
76	7.813	7.874	7.936	7.999	8.063	8.127	8.193	8.259	8.327	8.395
77	8.465	8.535	8.607	8.680	8.754	8.829	8.905	8.983	9.061	9.141
78	9.222	9.305	9.389	9.474	9.561	9.649	9.739	9.831	9.924	10.02
79	10.11	10.21	10.31	10.41	10.52	10.62	10.73	10.84	10.95	11.07
80	11.18	11.30	11.42	11.54	11.67	11.80	11.93	12.06	12.20	12.34
81	12.48	12.63	12.77	12.93	13.08	13.24	13.41	13.57	13.74	13.92
82	14.10	14.28	14.47	14.66	14.86	15.07	15.28	15.49	15.71	15.94
83	16.17	16.41	16.66	16.91	17.17	17.44	17.72	18.01	18.31	18.61
84	18.93	19.25	19.59	19.94	20.30	20.68	21.07	21.47	21.89	22.32
85	22.77	23.24	23.74	24.25	24.78	25.34	25.92	26.53	27.16	27.83
86	28.53	29.27	30.05	30.86	31.73	32.64	33.60	34.63	35.72	36.88
87	38.11	39.43	40.84	42.36	44.00	45.76	47.68	49.76	52.02	54.51

APPENDIX 8: PHYSICAL CONSTANTS AND CONVERSION FACTORS

Physical Constants

Avogadro constant	$N_A = 6.022 \times 10^{23}$ particles/mole
Boltzmann constant	$k = 8.617 \times 10^{-5}$ eV/K $= 1.38 \times 10^{-23}$ J/K
Charge on electron	$e = 1.602 \times 10^{-19}$ C
Planck constant	$h = 4.136 \times 10^{-15}$ eV · s $= 6.626 \times 10^{-34}$ J · s
Speed of light	$c = 2.998 \times 10^{8}$ m/s
Electron mass	$m = 9.1095 \times 10^{-31}$ kg

Useful Conversion Factors

1 nm $= 10^{-9}$ m $= 10$ Å $= 10^{-7}$ cm
1 eV $= 1.602 \times 10^{-19}$ J
0°C = 273 K
°C = (°F − 32) × 5/9
K = °C + 273

Prefixes

m = milli = 10^{-3}	k = kilo = 10^{3}
μ = micro = 10^{-6}	M = mega = 10^{6}
n = nano = 10^{-9}	G = giga = 10^{9}
p = pico = 10^{-12}	T = tera = 10^{12}

APPENDIX 9: JCPDS–ICDD CARD NUMBERS FOR SOME COMMON MATERIALS

Pearson symbol	Material Card Number						
cP1	α-Po	$AuTe_{1.7}$					
	19-899	19-528					
cP2	CsCl	β-CuZn	FeAl				
	5-607	2-1231	33-20				
cP4	Cu_3Au						
	35-1357						
cI2	Cr	Fe	Ta	W			
	6-694	6-696	4-788	4-806			
cF4	Ag	Al	Au	Cu	Ni	Pb	
	4-783	4-787	4-784	4-836	4-850	4-686	
cF8	AlN	CaO	GaAs	Ge	MgO	NaCl	NiO
	25-1495	37-1497	32-389	4-545	45-0946	5-628	4-835
	Si	TiC	TiN	ZnS			
	27-1402	32-1383	38-1420	5-566			
cF12	CaF_2						
	35-816						
cF16	Fe_3Al						
	45-1203						
hP2	Cd	Mg	Ti	Zn			
	5-674	35-821	44-1294	4-831			
hP4	AlN	PTFE	ZnO	ZnS			
	25-1133	27-1873	36-1451	36-1450			
hR10	α-Al_2O_3						
	46-1212						

APPENDIX 10: CRYSTAL STRUCTURES AND LATTICE PARAMETERS OF SOME SELECTED MATERIALS

Material	Symbol	Crystal structure	Pearson symbol	Lattice parameters		
				a	c	c/a
α-Alumina	α-Al_2O_3	corundum	hR10	0.4759	1.2993	2.7303
Aluminum	Al	fcc	cF8	0.4049		
Aluminum nitride	AlN	zinc blende	cF8	0.4120		
Aluminum nitride	AlN	wurtzite	hP4	0.3111	0.4979	1.6005
β-Brass	β-CuZn	CsCl	cP2	0.2948		
Cadmium	Cd	hcp	hP2	0.2979	0.5618	1.8857
Calcium oxide	CaO	NaCl	cF8	0.4811		
Cesium chloride	CsCl	CsCl	cP2	0.4123		
Chromium	Cr	bcc	cI2	0.2884		
Copper	Cu	fcc	cF4	0.3615		
Copper-gold	Cu_3Au	simple cubic	cP4	0.3749		
Calcium fluoride	CaF_2	fluorite	cF12	0.5463		
Gallium arsenide	GaAs	zinc blende	cF8	0.5654		
Germanium	Ge	diamond cubic	cF8	0.5658		
Gold	Au	fcc	cF4	0.4079		
Gold telluride	$AuTe_{1.7}$	simple cubic	cP1	0.2961		
α-Iron	α-Fe	bcc	cI2	0.2866		
Iron aluminide	FeAl	cubic	cP2	0.2895		
Iron aluminide	Fe_3Al	cubic	cF16	0.5793		
Lead	Pb	fcc	cF4	0.4951		
Magnesium	Mg	hcp	hP2	0.3209	0.5211	1.6238
Magnesium oxide	MgO	NaCl	cF8	0.4211		
Nickel	Ni	fcc	cF4	0.3524		
Nickel oxide	NiO	NaCl	cF8	0.4177		
α-Polonium	α-Po	simple cubic	cP1	0.3359		
Silicon	Si	diamond cubic	cF8	0.5431		
Silver	Ag	fcc	cF4	0.4086		
Sodium chloride	NaCl	NaCl	cF8	0.5640		
Tantalum	Ta	bcc	cI2	0.3306		
Titanium	Ti	hcp	hP2	0.2951	0.4683	1.5871
Titanium carbide	TiC	NaCl	cF8	0.4327		
Titanium nitride	TiN	NaCl	cF8	0.4242		
Tungsten	W	bcc	cI2	0.3165		
Zinc	Zn	hcp	hP2	0.2665	0.4947	1.8563
Zinc oxide	ZnO	wurtzite	hP4	0.3250	0.5207	1.6021
Zinc sulfide	ZnS	zinc blende	cF8	0.5406		
Zinc sulfide	ZnS	wurtzite	hP4	0.3821	0.6257	1.6376

Bibliography

Classic Books and Historical Perspectives

These books are particularly good for students interested in the development and history of the X-ray diffraction technique for characterization of materials.

Bijvoet, J. M., Burgers, W. G., and Hägg, G., Eds. (1969), *Early Papers on Diffraction of X-Rays*, published for the International Union of Crystallography by N. V. A. Ossthoek's Uitgeversmaatschappij, Utrecht, The Netherlands. Some of the classic papers, including those by Max von Laue and W. H. and W. L. Bragg, in x-ray diffraction are reprinted in this book. Should be of interest to students wanting to know about the history of x-ray diffraction.

Bragg, W. H. (1929), *An Introduction to Crystal Analysis*, Van Nostrand, New York. W. H. Bragg is the father, and this book is based on a series of lectures he gave in 1928 at University College in Aberystwyth, Wales.

Bragg, W. H., and Bragg, W. L. (1915), *X-Rays and Crystal Structure*, G. Bell and Sons, London, UK. The classic book by the Braggs (father and son). This is their first book about x-rays and their use in crystal structure analysis. Although it is over 80 years old, it is very readable and the descriptions are clear and concise.

Bragg, W. H., and Bragg, W. L., Eds., *The Crystalline State*, Macmillan, New York.
Volume 1: Bragg, W. L. (1934), *A General Survey*. A clear survey of x-ray diffraction and its use for crystal structure determination.
Volume 2: James, R. W. (1954), *The Optical Principles of the Diffraction of X-Rays*. Advanced theory of x-ray diffraction. A mathematical approach only for the very enthusiastic student.
Volume 3. Lipson H., and Cochran, W. (1953), *The Determination of Crystal Structures*. Advanced treatment of structure analysis from a theoretical point of view.

Ewald, P. P., Ed. (1962), *Fifty Years of X-Ray Diffraction*, published for the International Union of Crystallography by N. V. A. Ossthoek's Uitgeversmaatschappij, Utrecht, The Netherlands. Provides a very interesting and readable historical perspective of x-ray diffraction. Personal reminiscences by some of the big names in the development of the technique are particularly enjoyable.

X-Ray Safety

As we mentioned in Sec. 3.5, commercial x-ray diffractometers are designed to meet extremely stringent safety requirements. However, if you want to find out more about the effects of exposure to x-rays, we recommend that you consult the following publications:

Henry, H. F. (1969), *Fundamentals of Radiation Protection*, John Wiley and Sons, New York.

Lakey, J. R. A., and Lewins, J. D., Eds. (1987), *ALARA Principles, Practice and Consequences*, Adam Hilger, Bristol, UK. Proceedings of a conference about the principles of ALARA.

Principles and Applications of Collective Dose in Radiation Protection (1995), NCRP Report No. 121, National Council on Radiation Protection and Measurements, Bethesda, MD.

Radiation: Doses, Effects, Risks (1985), United Nations Environment Program. A short booklet explaining some of the naturally occurring sources of ionizing radiation and their effects on the human body.

Crystal Structure

To understand the principles of x-ray diffraction, it is important to have at least a basic understanding of crystal structures. For the interpretation of x-ray diffraction patterns obtained from crystalline materials, a knowledge of crystal structures is essential. To work through the experimental modules and exercises in this book all the information you need to know about crystal structures is provided in Chapter 2. For more information, at the introductory level, most standard undergraduate materials science and engineering textbooks have a chapter on crystal structures. For example, we recommend the following textbooks:

Callister, W. D. (1997), *Materials Science and Engineering: An Introduction*, 4th. ed., John Wiley and Sons, New York. Chapter 3 is on the structure of crystalline solids.

Smith, W. F. (1996), *Principles of Materials Science and Engineering*, 3rd. ed., McGraw-Hill, New York. Chapter 3 is on crystal structure and crystal geometry.

At a more advanced level you may find the following books useful:

Barrett C. S., and Massalski, T. B. (1980), *The Structure of Metals*, 3rd. ed., McGraw-Hill, New York. A classic book dealing with the structures of metals and structural characterization by x-ray diffraction.

Kingery, W. D., Bowen, H. K., and Uhlmann, D. R. (1976), *Introduction to Ceramics*, 2nd. ed., John Wiley and Sons, New York. Chapter 2 is an overview of the structures of ceramic materials.

Pearson, W. B. (1972), *The Crystal Chemistry and Physics of Metals and Alloys*, John Wiley and Sons, New York. Includes notation for crystal structures including Pearson symbols that we have used in our book. Also discusses crystal structures.

X-Ray Diffraction

The technique of x-ray diffraction has been the subject of many books. Here are some that we found useful when we were learning the technique:

Azaroff, L. V. (1968), *Elements of X-Ray Crystallography*, McGraw-Hill, New York. Provides a fundamental understanding of the basic principles of x-ray diffraction. Emphasizes the reciprocal lattice concept. Most useful for a course in physics rather than materials science.

Azaroff, L. V., and Buerger, M. J. (1958), *The Powder Method in X-Ray Crystallography*, McGraw-Hill, New York. One of the early books on recording and interpreting x-ray powder patterns.

Barrett, C. S., and Massalski, T. B. (1980), *The Structure of Metals*, 3rd. ed., McGraw-Hill, New York. A classic book dealing with the structures of metals and structural characterization by x-ray diffraction.

Cullity, B. D. (1978), *Elements of X-Ray Diffraction*, 2nd. ed., Addison-Wesley, Reading, MA. An introduction to x-ray diffraction. The background information on structures and symmetry makes it suitable for undergraduates and first-year graduate students in materials science.

Guinier, A. (1963), *X-Ray Diffraction*, W. H. Freeman, San Francisco, CA. A comprehensive book dealing with the x-ray diffraction of crystals, imperfect crystals, and amorphous solids. Determination of crystal structures is not covered.

Klug, H. P., and Alexander L. E. (1974), *X-Ray Diffraction Procedures*, 2nd. ed., John Wiley and Sons, New York. A very comprehensive textbook dealing with many aspects of powder diffraction (both theory and practice).

McLachlan, Jr., D. (1957), *X-Ray Crystal Structure*, McGraw-Hill, New York. A book more useful to physics graduate students or to those whose interests lie more in crystallography.

Nuffield, E. W. (1966), *X-Ray Diffraction Methods*, John Wiley and Sons, New York. A book dealing with all methods of x-ray diffraction, particularly directed at students in the earth sciences.

Schwartz, L. H., and Cohen, J. B. (1977), *Diffraction from Materials*, 2nd. ed., Academic Press, New York. Covers not only x-ray diffraction but also electron and neutron diffraction.

Stout, G. H., and Jensen, L. H. (1968), *X-Ray Structure Determination*, Macmillan, New York. An advanced book dealing with practical problems of solving crystal structures through x-ray diffraction.

Warren, B. E. (1969), *X-Ray Diffraction*, Addison-Wesley, Reading, MA. This is an excellent advanced textbook useful to graduate students in materials science and physics. The treatment of the subject is theoretical and therefore may be a bit heavy for undergraduate students.

Woolfson, M. M. (1977), *An Introduction to X-ray Crystallography*, 2nd. ed., Cambridge University Press, Cambridge, UK. A good book on crystallography and experimental details of recording x-ray diffraction patterns. Also lists computer programs to calculate structure factors, electron density maps, anomalous scattering, and solution of crystal structures.

Woolfson, M. M., and Fan, H-F. (1995), *Physical and Non-Physical Methods of Solving Crystal Structures*, Cambridge University Press, Cambridge, UK. A comprehensive book on the theory of solving crystal structures.

Data Books and Resources for Crystallographic Information

Here are some of the sources of information you might find useful when trying to index an x-ray diffraction pattern or to identify a material.

Ca.R.Ine. Crystallography 3.0, Distributed by ESM Software, 2234 Wade Court, Hamilton, OH 45013; Tel: 513-738-4773; FAX: 513-738-4407. Software programs to generate real lattices in 3D, reciprocal lattices in 3D and 2D, stereographic projections, and ideal diffraction patterns (peak positions and intensities). For a free demonstration of the software you can go to the following web site on the internet: http://www.lmcp.jussieu.fr/sincris/logiciel/carine/

Donnay, J. D. H., Donnay, G., Cox, E. G., Kennard, O., and Vernon, M. (1963), *Crystal Data Determinative Tables*, 2nd. ed., The American Crystallographic Association, Washington. Crystallographic data on many compounds.

Fang, J. H., and Bloss, F. D. (1966), *X-Ray Diffraction Tables*, Southern Illinois University Press, Carbondale, IL. Tables of interplanar spacings (d) corresponding to 2θ for various wavelengths. This information can now be obtained using most software packages available with modern x-ray diffractometers.

Lide, D. R., Ed. (1996–1997), *CRC Handbook of Chemistry and Physics*, 77th. ed., CRC Press, Boca Raton, FL. Often one of the first sources you should try when you want information about a material.

Massalski, T. B., Ed. (1990), *Binary Alloy Phase Diagrams*, 2nd. ed., ASM International, Materials Park, OH. Volume 1: Ac–Ag to Ca–Zn; Volume 2: Cd–Ce to Hf–Rb; Volume 3: Hf–Re to Zn–Zr. The standard reference source for phase diagrams of alloys.

Pearson, W. B., *A Handbook of Lattice Spacings and Structures of Metals and Alloys*, Vol. 1 (1958), Vol. 2 (1967), Pergamon Press, Oxford, UK. Lists lattice spacings of alloys and compounds as a function of solute content.

Phase Diagrams for Ceramists, The American Ceramic Society, Columbus, OH.
Volume 1: Levin, E. M., Robbins, C. R., and McMurdie, H. F., Eds. (1964).
Volume 2: Levin, E. M., Robbins, C. R., and McMurdie, H. F., Eds. (1969).
Volume 3: Levin, E. M., and McMurdie, H. F., Eds. (1975).
Volume 4: Roth, R. S., Negas, T., and Cook, L. P., Eds. (1981).
Volume 5: Roth, R. S., Negas, T., and Cook, L. P., Eds. (1983).
The standard reference source for phase diagrams of ceramics.

Powder Diffraction File, Inorganic Phases Search Manual (Hanawalt), International Centre for Diffraction Data, Swarthmore, PA. Tabulation of peak positions and intensities for inorganic materials. Essential for identification of unknown materials by x-ray diffraction. An updated volume is published each year.

Villars, P., and Calvert, L. D. (1991), *Pearson's Handbook of Crystallographic Data for Intermetallic Phases* (in four volumes), ASM International, Materials Park, OH. Crystallographic information on pure metals, semiconductors, and ceramic compounds, in addition to intermetallic phases.

Wilson, A. J. C., Ed. (1995), *International Tables for Crystallography,* Vol. C., *Mathematical, Physical and Chemical Tables,* Kluwer Academic Publishers, Dordrecht, The Netherlands. Covers many aspects of diffraction, including x-ray diffraction, and provides a comprehensive list of data on characteristic wavelengths, absorption coefficients, atomic scattering factors, etc. An invaluable reference.

Wyckoff, R. W. G. (1963), *Crystal Structures,* 2nd. ed., John Wiley and Sons, New York. Describes many crystal structures.

Index